Advanced
Modular
Mathematics

STATISTICS 1

Gerald Westover
Graham Smithers

**SECOND
EDITION**

COLLINS

nec
NATIONAL
EXTENSION
COLLEGE

Published by HarperCollins Publishers Limited
77–85 Fulham Palace Road
Hammersmith
London W6 8JB

www.**Collins**Education.com
On-line Support for Schools and Colleges

© National Extension College Trust Ltd 2000
First published 2000
ISBN 000 322521 6

This book was written by Gerald Westover and Graham Smithers for the National Extension
College Trust Ltd. Part of the material was originally written by Mik Wisnieski and Clifford Taylor.

British Library Cataloguing in Publication Data
A catalogue record for this publication is available from the British Library.

Original internal design: Derek Lee
Cover design and implementation: Terry Bambrook
Page layout: Eric Coles
Project Editor: Hugh Hillyard-Parker
Printed and bound: Martins the Printers Ltd., Berwick-upon-Tweed

The authors and publishers thank Dave Wilkins for his comments on this book.

The National Extension College is an educational trust and a registered charity
with a distinguished body of trustees. It is an independent, self-financing organisation.

Since it was established in 1963, NEC has pioneered the development of flexible
learning for adults. NEC is actively developing innovative materials and systems for
distance-learning options from basic skills and general education to degree and
professional training.

For further details of NEC resources that support Advanced Modular Mathematics,
and other NEC courses, contact NEC Customer Services:

National Extension College Trust Ltd
18 Brooklands Avenue
Cambridge CB2 2HN
Telephone 01223 316644, Fax 01223 313586

Y... ... ...ike to visit:

fireand**water**.com
...ook lover's website

UNIT

S1

Contents

S1

Advanced Modular Mathematics

This book is one of a series covering the Edexcel Advanced Subsidiary (AS) and Advanced GCE in Mathematics. It covers all the subject material for Statistics 1 (Unit S1), examined from 2001 onwards.

While this series of text books has been structured to match the Edexcel specification, we hope that the informal style of the text and approach to important concepts will encourage other readers whose final exams are from other Boards to use the books for extra reading and practice. In particular, we have included references to the OCR syllabus (see below).

This book is meant to be *used*: read the text, study the worked examples and work through the Practice questions and Summary exercises, which will give you practice in the basic skills you need for maths at this level. Many exercises, and worked examples, are based on applications of the mathematics in this book. There are many books for advanced mathematics, which include many more exercises: use this book to direct your studies, making use of as many other resources as you can.

There are many features in this book that you will find particularly useful:

- Each **section** covers one discrete area of the new Edexcel specification. The order of topics is exactly the same as in the specification.

- **Practice questions** are given at regular intervals throughout each section. The questions are graded to help you build up your mathematical skills gradually through the section. The **Answers** to these questions come at the end of the relevant section.

- **Summary exercises** are given at the end of each section; these include more full-blown, exam-type questions. Full, worked solutions are given in a separate **Solutions** section at the end of the book.

- In addition, we have provided a complete **Practice examination paper**, which you can use as a 'dummy run' of the actual exam when you reach the end of your studies on S1.

- Alongside most of the headings in this book you will see boxed references, such as: OCR **S2** 5.12.4(a) These are for students following the OCR specification and indicate which part of that specification the topic covers.

- Your work on this book will provide opportunities for gathering evidence towards Key Skills, especially in Communication and Application of Number. These opportunities are indicated by a 'key' icon,

  for example: **C** 3.2 (see Appendix 3 for more information).

The National Extension College has more experience of flexible-learning materials than any other body (see p. ii). This series is a distillation of that experience: Advanced Modular Mathematics helps to put you in control of your own learning.

Mathematical models in statistics

INTRODUCTION   In this section we shall introduce the idea of statistics and the reason for its existence. In particular we shall see how:

● a sample is necessary for predicting the views of a population

● the sample needs to be random if each member of the population is to have an equal chance of expressing its view

● the types of data considered can be characterised as either discrete, continuous or qualitative

● descriptive statistics describe the results obtained from a sample whereas inferential statistics use that sample to draw conclusions about the poulation as a whole.

The mathematics of uncertainty

In your study of mathematics so far, you will have already developed a number of skills and knowledge relating to a variety of mathematical applications. Virtually all of what you will have met so far relates to such mathematical principles under conditions of certainty. For example, when we examine a situation where $y = 3x^2$, we take it for granted that this relationship is known for certain and will not vary or change at random.

Mathematical statistics recognises that there are areas in mathematics where such relationships and conditions will not always apply – in other words where a degree of *uncertainty* exists. Consider the simple illustration of tossing a coin and noting which side of the coin shows. Clearly, this outcome is not known for certain. We cannot guarantee the result of our action.

Statistics as a subject area is concerned with allowing us to reach conclusions or make decisions in the face of such uncertainty. Naturally, we still wish to apply appropriate mathematical principles and logic to examine such situations. Indeed there is a dual focus throughout the module.

● On the one hand, we shall be *deriving certain key theoretical principles* of mathematical statistics.

● On the other hand, we shall be seeking to *apply these principles* and the theory to a variety of practical situations. Statistics is, above all, an applied discipline.

This section begins by examining the principles whereby we can describe the key features of a set of data that we have obtained. In later sections you will be introduced to a variety of statistical means of achieving this. You will move on to the important area of probability. This is the mechanism whereby we can introduce concepts of uncertainty into our mathematical principles.

Using statistical methods and techniques

Statistical methods and techniques are applicable to a tremendously diverse range of other disciplines. A few examples will illustrate just how worthwhile the study of statistics is, not just in itself, but also for its relevance to other areas of study.

Typical statistical applications that make use of the topics we shall be covering in this module include:

- Is there any connection between people's social and economic class and their tendency to vote for a particular political party?
- Has the recent TV advertising campaign undertaken by a high street bank been successful in attracting more customers?
- Have the changes introduced in education in terms of the National Curriculum improved student learning?
- Does the treatment for the HIV virus being tested in the laboratory reduce the mortality rate?
- Are more women in favour of legalised abortion than men?
- Has the introduction of lead-free petrol reduced the level of pollution in city centres?

We could extend this list of areas suitable for statistical analysis but those shown illustrate the role that statistics has to play across a range of different subject areas.

From your study of mathematics so far, you will appreciate that mathematics has its own vocabulary and terminology. Statistics is a specialised branch of mathematics and has developed its own vocabulary. In order to proceed, you need to understand the meaning of the more important terms. Further terms will be introduced as and when they are needed.

Populations and samples

OCR **S2** 5.12.4 (a)

In the process of collecting and analysing data we must usually distinguish between a sample and the population. (Note that the word 'population' in statistics does not necessarily refer to people.)

> A *population* in statistics refers to the entire set of data that exists.
>
> A *sample*, on the other hand, refers to a carefully chosen and representative part of the population.

This distinction between sample and population is an important one.

Assume that we wish to collect data on family characteristics (sizes, ages, sex, etc.) for the entire country. The population would, therefore, consist of all families in the UK at some moment in time.

Where a population is large or inaccessible, it is usual to take a sample from the population. We can use the information contained in the sample to make judgements or inferences about the whole population. Naturally we would try to ensure that the sample was representative of the larger group, that is, the population it was meant to represent.

Sometimes we have to use a sampling method because the process of testing destroys the sample. If the official taster in a chocolate bar factory tried every bar, there'd be none left to sell!

Practice questions A

1 Political opinion polls are often quoted in newspapers. What is a typical sample size?

2 You wish to find out what motorists think of the traffic situation. What would be a realistic sample size for you to choose?

Sampling methods

OCR **S2** 5.12.4 (b)

In order to obtain information about a population, we usually take a sample and perform certain calculations and tests and then extrapolate back to the population. Often a **random sample** is taken. A random sample is one in which every member of the population has an equal chance of appearing in the sample.

If it is possible to list and give a unique identification to each member of a population, then the list is referred to as a **sampling frame** and the existence of such a frame represents an ideal situation.

The list of members of the population can be mixed thoroughly and the sample drawn randomly. For example, if a survey were being carried out about the chocolate-eating habits of the members of a certain college, then a list could be fairly simply drawn up and names selected randomly for the sample, rather like in a lottery.

For larger populations, e.g. adults in the United Kingdom, it may be feasible to sample using the Electoral Register, which is effectively a sampling frame for the voting population. However, it will not include all adults in the United Kingdom and can never be completely up to date.

For surveys on large populations it is common practice to conduct a **pilot study** first to ensure that sensible responses are obtained and that all of the information required is being obtained.

Practice questions B

1 When pollsters stand on street corners seeking out people's opinions, are they taking a random sample?

2 When a telephone poll is taken, is this a random sample?

Variates or variables

We frequently use the term '**variable**' or '**variate**' when dealing with statistical data (the two terms are effectively interchangeable). Assume that we are conducting an investigation into the sizes of families in the UK. For each family in the investigation we could collect data on such things as:

- the number of people in the family group
- their ages
- their educational background
- which part of the country they live in

and so on. Each of these characteristics is a variable or variate in our investigation.

A variable will fall into one of three general categories. It may be:

- discrete
- continuous
- qualitative.

To explain each of three categories:

- A **discrete variable** is one which can only have certain numbers (or take certain fixed values). If the variable were the number of people in a family, it would be an example of a discrete variable. The number of people might be 1 or 2 or 3, for example, but it could not possibly be 1.4 or 5.3. Don't assume that a discrete variable can only take integer values. Shoe sizes, for example, are discrete – 7, $7\frac{1}{2}$, 8, $8\frac{1}{2}$, etc.

- A **continuous variable** is one which can take any numerical value within a given range. Imagine that we want to extend our investigation of families to look at diet patterns: we might want to analyse the weight of family members. This would be an example of a continuous variable, since, technically, we could measure this variable to any degree of accuracy that we wanted. We might, for example, measure the variable to the nearest kilogram, to the nearest gram, or to the nearest milligram.

- **Qualitative variables** are those that cannot be shown in terms of numbers. For example, suppose our chosen variable was the sex of the family members. We cannot express this variable sensibly as a number – we would probably record the data collected as 'male' or 'female'.

Practice questions C

1 You intend to carry out a survey concerning telelvision viewing habits. Give examples of discrete, continuous and qualitative variables that you might consider in this context.

2 How long can you hold your breath? What sort of variable is this?

Descriptive and inferential statistics

The final piece of vocabulary that we need to introduce at this stage relates to the difference between descriptive and inferential statistics.

- **Descriptive statistics** as a subject is concerned with the techniques for collecting, grouping, summarising and presenting data: techniques which allow us to describe the main features of a set of data.

Returning to our sample of families, we might use such techniques to describe statistically the number of families in the sample that consist entirely of people over the age of 60.

- **Inferential statistics** is concerned with trying to reach a conclusion on a population when only a sample is available. Typically, inferential statistics is concerned with reaching conclusions about some characteristic of the population based on the descriptive statistics of the sample.

  So, for example, we would try and determine how many families in the entire statistical population consisted of people over the age of 60, based only on the sample.

Inferential statistical methods become important because it is the population that we normally wish to examine. However, as it is usually only a sample for which we actually have data, we need to be able to move from sample to population with a reasonable degree of certainty.

As an illustration of the general processes involved in a statistical investigation, consider the following example.

Example

A manufacturer of batteries claims that his particular brand will remain usable for at least 1000 hours.

It is decided to check his claim and an independent observer takes a sample from his production line and conducts an experiment. He runs the batteries until they are dead and notes the time for each. This is a random or unpredictable quantity – an example of a continuous variable.

Having collected his data, he will then perform some calculations and produce some graphical illustrations of his results.

Finally, he will use his results to try to draw some conclusions about the manufacturer's claim – he will test a hypothesis about its validity.

This procedure is summarised in Fig. 1.1.

Figure 1.1

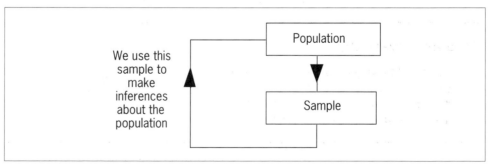

Practice questions D

 **N** 3.1,3.2,3.3

1 How might descriptive and inferential statistics be involved when considering the distribution of vowels in the English language?

2 Car registrations in the UK provide particular information about the car. Describe how you could use descriptive and inferential statistics to examine car registrations.

SUMMARY EXERCISE

1 What reasons can you think of which would prohibit the collection of all data for all families in the UK?

2 (a) Explain briefly:

   (i) why it is often desirable to take samples

   (ii) what you understand by a sampling frame.

   (b) Give an example of a sample frame suitable for use in a survey of attitudes of pupils in a school to a proposal to start the school day 15 minutes earlier.

3 For each of the following statistical variables, state whether it is discrete, continuous or qualitative:

   (a) your height

   (b) the colour of your hair

   (c) the number of people in your family group

   (d) the number of exam passes you have already

   (e) the time taken for the 08.15 from Liverpool Street Station to reach Chelmsford

   (f) the weight of a bag of flour.

SUMMARY

In this section we have seen that:

● a **population** in statistics refers to the entire set of data

● a **sample** in statistics refers to a carefully chosen representative part of the population

● a **random sample** is one in which every member of the population has an equal chance of being chosen.

● a **variable** is the term applied to the data we are considering

   – *discrete* variables, e.g. number of cars owned, shoe size

   – *continuous* variables, e.g. weight, length or time

   – *qualitative* variables, e.g. sex or colour.

● **descriptive** statistics describe the results obtained from a sample

● **inferential** statistics make inferences from the sample about the population.

ANSWERS

Practice questions A

1 Usually about 1000.

2 Probably no more than 100 – it would be too time-consuming to consider a larger sample.

Practice questions B

1 No, because not every person has an equal chance of being chosen, e.g. interviewers may be reluctant to question certain types of pedestrians.

2 It is certainly not a random sample of the population at large, e.g. not everybody has access to a phone or, indeed, will answer the phone.

Practice questions C

1 (a) Discrete – What is the number of different programmes viewed each day?

   (b) Continuous – What is the total time spent viewing on Monday?

   (c) Qualitative – Are the programmes good, bad or mixed? Which channel is watched most frequently?

2 Time is a continuous variable.

Practice questions D

1 (a) Take a sample of English prose and count the number of A, E, I, O and U's.
That will provide a descriptive statistic.

 (b) Use this sample to predict the most popular vowel in the English language.
That will provide an inferential statistic.

2 (a) Take a sample and produce a chart showing the number of different types of registrations.
That will provide a descriptive statistic.

 (b) Use the sample to predict how many cars on the road are at least three years old. That will provide an inferential statistic.

2

Representing data I:
Statistical diagrams

Having taken our sample, we will now look at ways of representing the data diagrammatically. First we will set up a frequency table and from this we will be able to draw a histogram and a frequency polygon. A cumulative frequency polygon will enable us to estimate useful statistics such as percentiles and quartiles. From this we will be able to draw a box and whisker diagram which will summarise diagrammatically all that we have found.

Should the data be discrete, a stem and leaf diagram would provide a neat diagrammatic summary.

Frequency tables

OCR **S1** 5.11.1 (a),(b)

 C 3.2

We shall illustrate and develop the appropriate statistical techniques with reference to the two data sets shown below in Tables 2.1 and 2.2.

Table 2.1	Height of 50 adult males, measured in cm								
168.3	169.1	175.9	178.8	172.3	182.9	173.9	184.1	168.9	178.1
175.2	167.7	169.3	189.8	179.7	187.9	160.6	171.3	159.8	167.2
195.3	166.5	164.8	173.3	175.0	178.1	172.3	172.5	183.5	171.0
163.0	169.1	172.4	182.4	172.3	172.1	180.1	175.2	183.4	170.5
175.0	184.2	183.2	176.2	171.5	175.0	178.1	175.1	181.0	178.1

Table 2.2	Heights of 50 adult females, measured in cm								
168.3	168.8	170.8	166.2	169.8	170.1	171.6	171.6	153.3	161.1
172.1	171.3	172.1	163.4	171.8	163.2	168.5	169.0	169.1	175.2
154.8	159.9	181.9	155.8	166.2	156.1	169.6	163.7	171.3	173.9
172.8	159.9	166.2	170.6	172.4	169.8	167.3	177.4	159.6	170.3
172.9	162.8	162.1	159.4	161.9	157.1	161.7	164.7	162.2	169.8

We shall be using these data sets to illustrate a number of key statistical concepts in this and subsequent sections, so it is worthwhile describing them in detail.

We have taken a sample of 50 adult males and 50 adult females and measured their heights (in centimetres, correct to one decimal place). We can easily imagine a study that we may be involved in where we have to compare heights of males and females in order to ascertain similarities and differences between the two sexes. The study might be for medical or health purposes, for some business organisation, for the police or armed forces and so on.

With the data in the form given, it is difficult to draw any conclusions about the heights of adults.

We cannot readily identify the key features of the two data sets. For the data to be useful it has to be put into a more manageable form. The first step in this process is often to construct a **frequency table**.

A frequency table for males is shown in Table 2.3, where the letter x is used to stand for the variable 'height' and the letter f is used to stand for the 'frequency'.

Table 2.3	Frequency table for male heights (cm)
height (x)	frequency (f)
155–	1
160–	3
165–	8
170–	12
175–	14
180–	9
185–	2
190–	0
195–	1
200–	0
Total frequency	50

Here the interval '155–', for example, is to be interpreted as $155 \leq x < 160$.

Note that an extra interval '200–' is included at the end to indicate that the previous interval stops at 200 and is therefore the same width as the earlier intervals.

We can now pick out some general features of the distribution:

● No one is less than 155 cm.

● No one is above 200 cm.

● Most males are between 170 and 180 cm.

Stages in constructing a frequency table

Step 1: From the raw data, determine the minimum and maximum values.

The difference between the two is known as the *range*. This allows us to see the range of data the table will have to deal with.

The lowest height in Table 2.1 is 159.8 and the largest is 195.3, so the table must encompass a range of about 40 cm. (As with many statistical methods it is useful to round numbers to some manageable and meaningful figure.)

Step 2: Choose the number of intervals to be shown in the table.

There is no absolute rule that can be applied, but we use our own common sense and, through trial and error, produce a table which looks appropriate for the data. It is conventional to have between 5 and 15 intervals in such a table, with fewer intervals for small data sets than for large. If there are too few intervals then important details about the data set may be lost as we aggregate the data. On the other hand, with too many intervals the data may be insufficiently aggregated and no patterns in the data will be apparent.

Step 3: Decide how large each interval should be.

In Table 2.3 the intervals are all 5 cm. Steps 2 and 3 are related and we must choose interval sizes and the number of intervals together. If possible, we should choose intervals which are all the same width. Here, a choice of 5 cm provides us with a sensible interval size and also gives nine intervals, which seems appropriate.

Step 4: Ensure that the boundaries of the intervals are clear and unambiguous.

In Table 2.3 the intervals are expressed so that there is no possible misunderstanding. If we had shown the intervals as:

155 to 160 cm, 160 to 165 cm, 165 to 170 cm, etc.

then when we aggregate the raw data we are uncertain whether to place an observation of, say, 160 in the first interval or in the second.

Step 5: Work through the data set allocating each observation to the appropriate interval.

Class boundaries

The data in Tables 2.1 and 2.2 was given correct to one decimal place. The first item in Table 2.1, namely 168.3, could therefore have been a measurement anywhere in the interval $168.25 \le x < 168.35$ since any measurement within this interval would round to 168.3 to one decimal place.

When the data is grouped into the frequency table, Table 2.3, the true limits of the intervals are therefore slightly different. The true limits of each interval are called the **class boundaries**.

As an example, consider the class 160– (i.e. $160 \le x < 165$). Strictly speaking, any height in the interval $159.95 \le x < 164.95$ would end up in this class and therefore these extremes determine the class more precisely.

Practice questions A

1 A sample of rats was weighed. The smallest weighed 102g whereas the largest weighed 1.1 kg. Suggest suitable class intervals in the case of:

(a) a sample of 20 rats

(b) a sample of 200 rats.

2 The lengths of a sample of worms were measured and the following class intervals were used:

1 cm $\le x < 2$cm, 2 cm $\le x < 3$ cm, …

Worm A had length 2.98 cm, worm B had length 1.04 cm and worm C had length 2 cm.

In which intervals would their lengths be recorded?

Other types of frequency table   OCR **S1** 5.11.1 (b),(e)

It is quite common, and frequently useful, to show a frequency distribution in terms of **relative** and **cumulative** *frequencies* as well as in absolute terms. Table 2.4 shows these for males.

Table 2.4		Frequency table for males showing relative and cumulative frequencies	

Heights (cm) x	Frequency f	Relative frequency	Cumulative frequency
155–	1	0.02	1
160–	3	0.06	4
165–	8	0.16	12
170–	12	0.24	24
175–	14	0.28	38
180–	9	0.18	47
185–	2	0.04	49
190–	0	0.00	49
195–	1	0.02	50
200–	0	0.00	50
Total	50	1.00	50

- The cumulative frequencies show the number of observations *up to and including that interval*; so, for example, there were 24 males who had heights up to, but not including, 175 cm.

- Relative frequencies are useful for comparing distributions of data with probability distributions. They are simply calculated as:

  frequency ÷ total frequency

  So, for example, for the interval 160– with frequency $f = 3$, the relative frequency is $3 ÷ 50 = 0.06$.

  Note that relative frequencies should always add up to 1 (with allowances made for rounding).

Practice questions B

1   Complete the following table:

Height (cm)	Frequency	Relative frequency	Cumulative frequency
50–	5	0.1	5
55–	10		
60–	19		
65–	9		
70–	3		
75–	2		
80–	2		
Total	50		

2   Complete the following table:

Height (cm)	Frequency	Cumulative frequency
80–		3
90–		8
100–		19
110–		36
120–		40

Histograms

OCR **S1** 5.11.1 (b),(c)

N 3.1,3.2,3.3

Whilst frequency tables are a useful first step in describing some set of data, a visual presentation of the frequency table can be even more useful. The diagram of a frequency table is known as a **histogram**. Figure 2.1 shows a histogram for the height of males shown in Table 2.3.

Figure 2.1

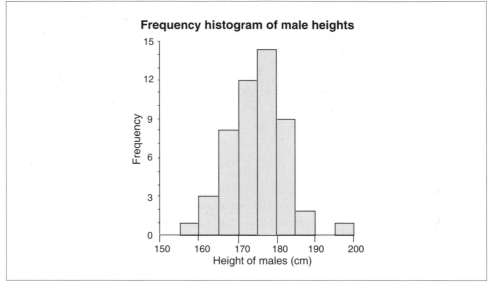

As you can see the histogram provides a clear and quickly understood picture of the pattern of the variable. We gain an immediate overall impression of the frequency distribution from the histogram.

Histograms are quite easy to construct. The variable we are analysing – height of males – is shown on the horizontal axis and the frequencies are shown on the vertical axis. Naturally, we must exercise all the usual caution when drawing such a diagram: ensuring that the scales are appropriate, titles and labels are clearly shown and so on. Note also that a histogram has the columns joined together, corresponding to the continuous nature of the data being illustrated. Where there is a 'gap' between columns, as there is in Fig. 2.1, this is not a gap at all, rather an interval with zero frequency.

Practice questions C

1   Use the following histogram to complete the table alongside.

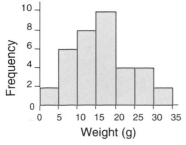

Weight (g)	Frequency	Cumulative frequency
0–	2	2
5–		
10–		
15–		
20–		
25–		
30–		
35–		

2   Refer to question 1 above. How many weigh 15g or more?

Problems met in constructing a histogram

Constructing a histogram from a frequency table is usually straightforward, but there are two aspects that we need to be aware of.

Open-ended intervals

These are intervals which have no upper (or lower) limit specified. In Table 2.3 assume that we had one interval where heights were defined as less than 160 cm rather than 155– or $155 \leq x < 160$. This would be an open-ended interval given that we had not specified an exact lower limit.

Equally, we could have an upper open-ended interval. This causes some difficulty when drawing the histogram as we do not know where to draw the lower (or upper) limit for the appropriate bar in the diagram. Typically we might use such open-ended intervals if we had relatively few observations scattered across several intervals at one end of the distribution.

Conventionally, we can deal with such open-ended intervals in one of two ways.

- We could give this interval the same width as the interval next to it. Here we would assume a lower limit of 155 for this open-ended interval so that this gives us an interval which has the same width (of 5 cm) as all our other intervals.

- There may be a logical limit implied in the data. Assume, for example, that we had constructed a frequency table showing the distribution of the ages of a sample of people. One group might be those who are 'under 5 years old'. We could realistically assume a lower limit of 0 for such an interval.

Unequal intervals

The second potential problem occurs when we have a frequency table where the intervals are not all of the same width. Because intervals may be of different widths, we need to compare not only the height of each bar but also its width relative to the others. In other words, in a histogram we are actually comparing areas (Height × Width = Area). To illustrate, let us examine Table 2.5.

Table 2.5	Frequency table for males (amended)
Height (cm) x	Frequency f
155–	1
160–	3
165–	8
170–	12
175–	14
180–	9
185–	3
200–	0
Total	50

Table 2.5 shows male heights but with three intervals from Table 2.3 amalgamated into one (185–, 190– and 195–).

In order to avoid gaining a distorted view of the last interval it is clear that in drawing the histogram we must adjust the diagram to allow for the extra width of this interval; we need to scale the frequency of that interval *downward* to bring it into line with our standard interval width. Conversely, if an interval were smaller than usual, we would scale the frequency upwards. To achieve this, we re-label the vertical axis as 'frequency density' and decide the height of each column as $\dfrac{\text{frequency}}{\text{class width}}$. This gives us frequency densities of 0.2, 0.6, 1.6, 2.4, 2.8, 1.8 and 0.2.

Figure 2.2

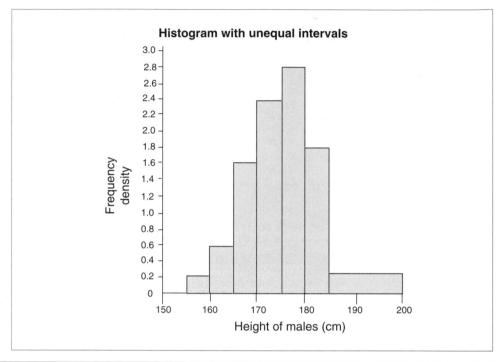

Histogram with unequal intervals

Example

Find the areas of the intervals $160 \le x < 165$ and $185 \le x < 200$.

Solution

Interval	Width	Height	Area
$160 \le x < 165$	5	0.6	3
$185 \le x < 200$	15	0.2	3

And these areas give the correct corresponding frequencies.

Practice questions D

1 Complete the following table:

Weight (g)	Frequency	Frequency density
$20 \le x < 30$	48	
$30 \le x < 40$	53	
$40 \le x < 60$	88	
$60 \le x < 90$	66	
$90 \le x < 100$	19	

2 Study the histogram below.

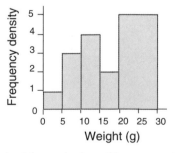

What size of sample does this represent?

3 An evening class in GCSE maths is mainly intended for people who have recently taken GCSE at school and want a better grade, but it is attended by a wider selection of people than that. On the right is a summary of their ages.

Draw a histogram for this data.

Age (x)	Frequency (f)
15–16	2
17–19	10
20–25	6
26–35	7

Histograms in relative form

Histograms may also be constructed to show relative, rather than absolute, frequencies. This is often useful where we wish to compare two or more data sets of differing sizes. If the sample of males had been of 500 whilst that of females was 350 then a histogram of absolute frequencies would be unhelpful, because there were more males than females in the data set. A histogram of *relative frequencies* would make comparison and analysis easier. The only difference in its construction is that the vertical scale will now run from 0 to 1 (or from 0 to 100 if we are showing percentages).

Frequency polygons and frequency curves

OCR **S1** 5.11.1 (a),(b)

We may construct a **frequency polygon** from a frequency table. We join the *midpoint* of each interval (i.e. the midpoint of the top of each column) together with a straight line. The appropriate diagram for heights of males is shown in Fig. 2.3.

Figure 2.3

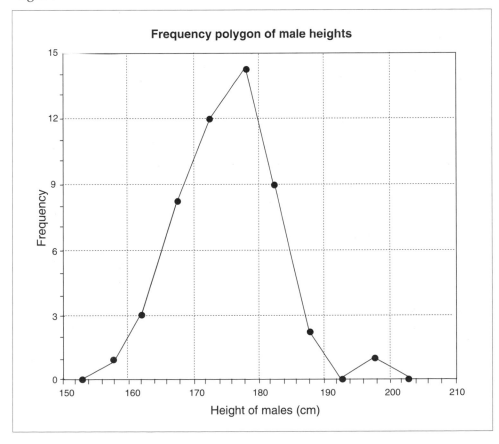

Each point plotted represents the *midpoint* of an interval. Note that at each end of the polygon we have plotted a point on the horizontal axis representing where the midpoint of the next interval would have been.

The frequency polygon is particularly useful for showing the general shape of the distribution and for comparing two or more such distributions on the same diagram.

If we now draw a similar diagram for heights of females superimposed on Fig. 2.3 we obtain Fig. 2.4 where comparison between the two is now much simpler.

Figure 2.4

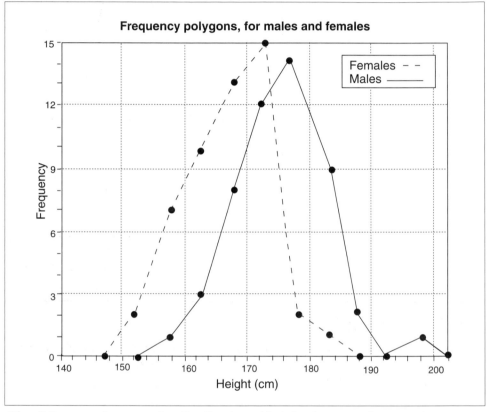

The differences between the distribution of heights between the sexes is now immediately apparent.

Frequency curves

The data that we have analysed has been sample data, and the frequency polygon is technically the shape of the distribution of the sample data. Frequently we may wish to infer the general shape of the population distribution. We achieve this by constructing a **frequency curve** rather than a polygon. This is done by smoothing the polygon.

The logic behind this is simple. Because we are dealing with sample data, we may encounter slight discrepancies in the sample distribution which we would not expect in the population. A smoothing process removes these.

Let us return to Fig. 2.3 to illustrate the process. If we examine the right-hand side of the polygon we see that it dips to zero for the interval $190 \leq x < 195$ and then rises again in the successive interval. It would be illogical to suppose that in the statistical population there was no one in this height group, but that there were people in the next. The plausible reason for such a dip lies with the fact that we have only a sample of the population. We can realistically assume there to be a slight discrepancy between the observed sample data and the expected population profile. The frequency curve smooths the dip out, as in Fig. 2.5.

Figure 2.5

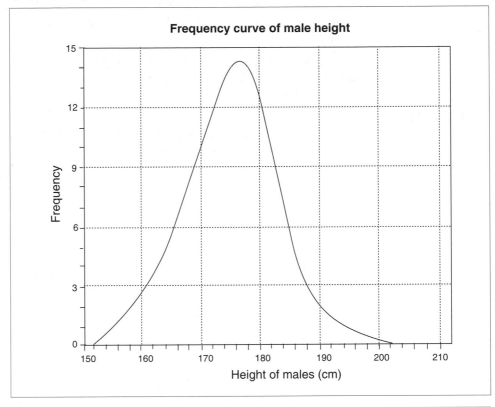

Practice questions E

1 A frequency polygon is required for the following data. Complete the necessary column. (You need not draw the polygon.)

Length (cm)	Frequency	Midpoints
10–	5	
20–	17	
30–	32	
40–	14	
50–	3	

2 Write down the midpoints of the following intervals:

(a) $18 \leq x < 24$

(b) $17 \leq x < 28$

(c) $8.1 \leq x < 9.3$

(d) $5.01 \leq x < 5.91$

Cumulative frequency diagrams

OCR **S1** 5.11.1 (b),(c)

The **cumulative frequency polygon** is drawn by plotting the cumulative frequency against the upper *end* of the interval (along the horizontal axis). The points are then joined together with straight lines. Figure 2.6 shows the cumulative frequency polygon for males. (Compare this with the cumulative frequency table for males, shown in Table 2.4.)

Figure 2.6

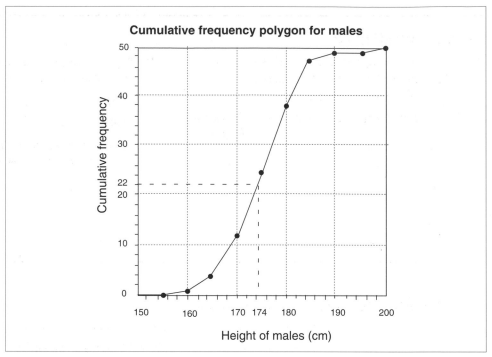

From the cumulative frequency polygon we can easily estimate the number of observations falling below (or above) a particular value.

For example, suppose that we wished to estimate the proportion of males with a height less than 174 cm. Given that this value is not consistent with the intervals we have used we must estimate from the cumulative frequency polygon. We see that approximately 44% (22/50) males are below this height.

If we now draw the cumulative frequency polygon for female heights superimposed on Fig. 2.6, we obtain Fig. 2.7 and we can read from this that approximately 88% (44/50) of females are below a height of 174 cm, for example.

Figure 2.7

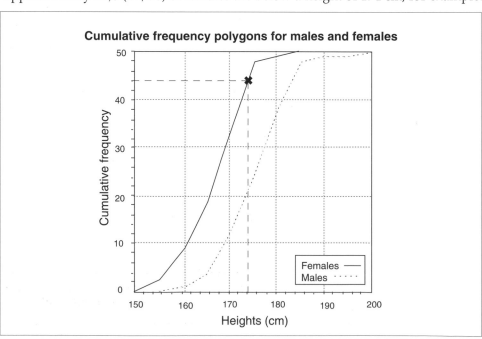

Practice questions F

1 A cumulative polygon is required for the following data. Complete the necessary columns. (You need not draw the cumulative polygon.)

Length (cm)	Frequency	End point	Cumulative frequency
5–	5		
7–	17		
9–	23		
11–	14		
13–	9		
15–	2		

2 It is required to represent the following data with a cumulative polygon.

Width (m):	6–	10–	14–	18–	22–26
Frequency:	9	17	23	14	3

Set up the necessary table. (You need not draw the cumulative polygon.)

Percentiles, deciles and quartiles

OCR **S1** 5.11.1 (b),(c)

Cumulative frequency polygons may also be constructed to show percentage frequencies. Such diagrams would show the percentage of observations falling below a specified value. It is useful when comparing two or more data sets where the total frequencies differ.

Such percentage cumulative frequency polygons also allow us to determine **percentile**, **decile** and **quartile** values of a data set. To see what these mean let us examine Fig. 2.8 which shows the cumulative percentage frequencies for males.

Figure 2.8

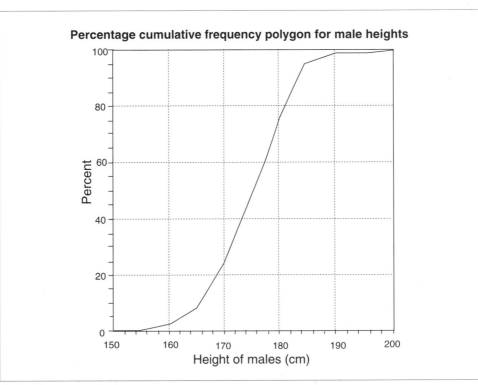

Percentage cumulative frequency polygon for male heights

- **Percentile** – A percentile is one of the values that divides the distribution into 100 equal parts. For example, we might find the fifth percentile. This would be the value for the variable such that 5% of observations fell below this value and 95% above. The 95th percentile would be the reverse of this: 95% of observations would fall below and 5% above.

- **Decile** – Where the chosen percentile coincides with multiples of 10 (10%, 20%, 30%, etc) then the value is known instead as a decile, e.g. the 30th percentile is called the 3rd decile.

- **Quartile** – Finally, where the data is divided not into percentages or into tenths, but rather into quarters, we refer to the quartiles (corresponding to the 25%, 50% and 75% points of the percentage cumulative frequency). The second quartile (Q_2) is the median. The other two (Q_1 and Q_3) are known as the **upper** and **lower quartiles.**

Example	Using the cumulative frequency polygon of Fig. 2.8 find the following:

(a)   the fifth percentile   (c)   the first decile (10%)

(b)   the 95th percentile   (d)   the upper quartile (75%)

Solution	The approximate answers from the diagram are:

(a)   162 cm   (c)   166 cm

(b)   184 cm   (d)   180 cm

The interpretation of these values is straightforward. The first decile, for example, indicates that 10% of adult males are below a height of 166 cm (and conversely 90% are above this height). Naturally such a statement is strictly true only for the sample data we are analysing; but, if we are confident that the sample is representative of the statistical population, we could reasonably infer that the same would be true of all adult males. Equally, we could use such statistics to compare adult males with adult females. You may wish to do this by yourself as a final exercise for this section.

Practice questions G

1   Copy and complete the following table:

Weight (g)	Frequency	End point	Cumulative frequency	Percentage cumulative frequency
10–	5			
20–	8			
30–	17			
40–	12			
50–	5			
60–70	3			

Now draw a percentage cumulative frequency polygon to represent this data and hence estimate:

(a) 20th percentile   (b) lower quartile   (c) median   (d) upper quartile.

Stem and leaf diagrams

OCR **S1** 5.11.1 (b),(c)

Stem and leaf diagrams provide a convenient way of representing discrete data. Suppose, for example, that we asked a group of 20 students how many TV programmes they had watched during the previous week.
Let's suppose their answers were:

39, 11, 12, 24, 25, 44, 8, 28, 15, 26, 32, 21, 5, 12, 24, 28, 13, 21, 15, 28

A neat way of presenting these results diagrammatically is to draw a **stem and leaf diagram**.

Begin by drawing a vertical line and, on the left hand side, mark the rows 0 (for single figures), 1 (for tens), 2 (for twenties), 3 (for thirties) and so on.

```
0 |
1 |
2 |
3 |
4 |
```

Now place the twenty readings in the appropriate row:

```
0 |  8   5
1 | 11  12  15  12  13  15
2 | 24  25  28  26  21  24  28  21  28
3 | 39  32
4 | 44
```

Finally, rearrange the numbers so that they are in order and only the second digit is recorded.

```
0 | 5  8
1 | 1  2  2  3  5  5
2 | 1  1  4  4  5  6  8  8  8
3 | 2  9
4 | 4
```

(For example, each 8 in row 2 stands for 28 and the 9 in row 3 stands for 39.)

This stem and leaf diagram shows very neatly how the figures are spread out.

Practice questions H

1   The number of TV programmes watched by a group of senior citizens during the last week was as follows:

22, 22, 18, 19, 21, 20, 39, 5, 9, 11, 25, 17, 15, 13, 11, 21, 18, 31, 7, 11.

(a) Represent the data with a stem and leaf diagram.   **C** 3.2

(b) What comparisons can you make with the TV viewing habits of the 20 students previously considered?

2   A stem and leaf diagram showing the number of letters received by a company during the last few weeks is given below:

```
0 | 0 0 1 1 2 4 6 7 8 9
1 | 2 2 4 5 5 5 5 5 5 5 7 8 9
2 | 1 4 7 7
3 | 2 2
4 | 8
```

(a) For how many days were the letters counted?

(b) What was the most common number of letters received per day?

Box and whisker plots

OCR **S1** 5.11.1 (b),(c)

Box and whisker plots provide a convenient way of representing the range, together with the median and quartiles. They can be used for both continuous and discrete data.

For example, look again at Fig. 2.8. From this we can see that:

- The minimum reading is 155 cm and the maximum is 200 cm.

- The lower quartile is approximately 170 cm.

- The median is approximately 176 cm.

- The upper quartile is approximately 180 cm.

A neat way of presenting these results diagrammatically is to draw a box and whisker plot.

Begin by drawing a vertical line and, having chosen a suitable scale, mark in the median and quartiles.

Figure 2.9

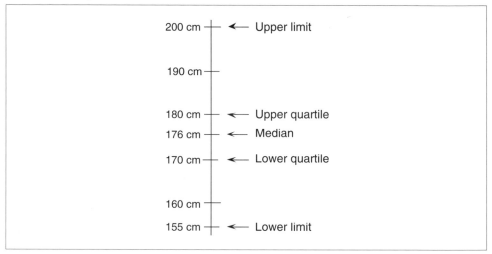

Now draw a box going from the lower to the upper quartile and mark in the median with a line across the box. This box represents the middle half of the distribution. Now draw lines out from the box to the extreme values. These lines (or 'whiskers') represent the lower and upper quarters of the distribution. Their chief purpose is to show the positions of the extreme values.

Figure 2.10

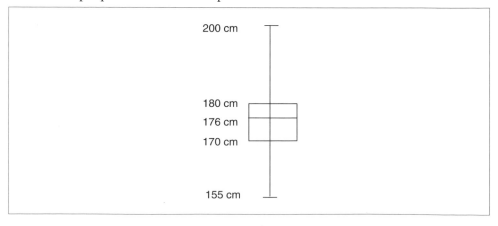

The final diagram is called a box and whisker plot. It shows very neatly how the figures are spread out and how the middle 50% are clustered between 170 cm and 180 cm.

Suppose we were given the box and whisker plot in Fig. 2.11.

Figure 2.11

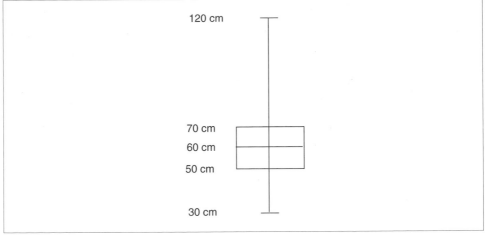

How might we interpret this summarised version of what may have been a large data set? Firstly, most of the data is concentrated around the interval 50 cm to 70 cm (i.e. 50% is included between the upper and lower quartiles). Secondly the range is 120–30 = 90 cm.

However the data clearly does not lie symmetrically within this range. The fact that the median is halfway between the lower and upper quartiles does suggest symmetry, however. We would be forced to conclude in this case that there are one or more items of data in the set which are extreme values and are atypical of the data set in general. Such atypical members of a data set are called **outliers**, and a box and whisker plot correctly interpreted can provide evidence of these.

Practice questions I

1 Weights of 100 cubes of sugar correct to nearest 5g:

Weight (g):	5	10	15	20	25
Frequency:	7	18	37	29	9

(a) Copy out and complete the table below:

Weight (g):	$2.5 \leq w < 7.5$	$7.5 \leq w < 12.5$	etc.
Frequency:	7	18	
Cum. freq:	7	25	
End of interval:	7.5	12.5	

(b) Use the table above to draw a cumulative polygon. Hence estimate:

(i) the lower quartile
(ii) median
(iii) upper quartile.

(c) Illustrate the data with a box and whisker plot.

2 Salaries of 100 managers in thousands of £:

Salaries:	≤ 10	≤ 20	≤ 30	≤ 40	≤ 60	≤ 80
Cum. freq:	11	35	72	80	88	100

(a) Illustrate the above figures with a cumulative polygon and hence estimate:

(i) the lower quartile
(ii) median
(iii) upper quartile.

(b) Illustrate the data with a box and whisker plot.

3 A sample of males and females were asked to record how many magazines they had bought during the year. The back-to-back stem and leaf diagram shown below illustrates the results:

Females							Males								
						5	1 1 3 4								**0**
					8	3	0 1 1 2 5 7 8								**1**
		7	6	6	5	2	3 4 4 5 8 9								**2**
8	8 7 5 5 2 1	0	0 1 2 1												**3**
			-8	7	5		3								**4**
				3	2										**5**
					1										**6**

(a) How many were interviewed in each sample?

(b) What is the greatest number of magazines read by any female in this sample?

(c) How many (i) females (ii) males read more than 30 magazines a year?

(d) Compare briefly the magazine reading habits of these females and males.

SUMMARY EXERCISE

1 Produce a frequency table for Table 2.2 using the intervals given below:

Heights, cm	Number of females
150–	
155–	
160–	
165–	
170–	
175–	
180–	
185–	
Total frequency	50

2 Construct a table like Table 2.4 for females.

3 Construct a comparable histogram for females, using your frequency table from Exercise 1.

4 Draw stem and leaf diagrams to represent the following data:

(a) 2, 6, 12, 12, 15, 18, 20, 20, 27, 31, 38, 42

(b) 4.8, 1.8, 3.0, 3.4, 1.2, 0.9, 3.3, 2.8, 2.3, 0.5, 2.2, 2.9, 3.7, 2.1, 4.2, 1.0, 2.5, 1.5, 1.6, 3.2

5 Using Fig. 2.7, estimate the lowest and highest values, quartiles and median for females. Hence represent the data with a box and whisker plot. (You should use the same scales as for the males on page 22.) Comment on any difference between the male and female distributions.

6 Telephone calls arriving at a switchboard are answered by the telephonist.

The following table shows the time, to the nearest second, recorded as being taken by the telephonist to answer the calls received during one day.

Time to answer (to nearest second)	Number of calls
10–19	20
20–24	20
25–29	15
30	14
31–34	16
35–39	10
40–59	10

Represent these data by a histogram.

Give a reason to justify the use of a histogram to represent these data.

7 At a health centre, where all the consultations are by appointment, a survey of 300 such appointments revealed that 265 were delayed. The times, in minutes, of these delays are summarised in the following table.

Delay, x minutes	Number of appointments
$0 < x < 1$	34
$1 \le x < 3$	50
$3 \le x < 5$	36
$5 \le x < 10$	65
$10 \le x < 15$	45
$15 \le x < 20$	15
$20 \le x < 30$	20

The illustration below was used in the report of the survey.

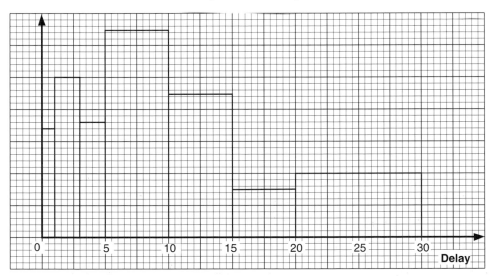

C 3.2

(a) Criticise the construction and presentation of this illustration.

(b) Draw your own histogram to represent these data and state any different impressions it displays from those in the illustration above.

The report also stated that, when an appointment is delayed, the median delay is 6.0 minutes, but that 50% of all appointments experience delays of less than 4.7 minutes.

(c) Show, by appropriate calculation, why **both** these statements are correct.

[AEB 1991]

8   The following table is extracted from a census report. It shows the age distribution of the population present on census night in Copeland, an area of Cumbria.

Population aged					
0–4	5–15	16–24	25–44	45–74	75 & over
4462	12 214	10 898	19 309	22 820	3364

Illustrate the data by means of a histogram. Make a suitable assumption about the upper bound of the class '75 and over'.

[AEB 1994]

SUMMARY

In this section we have seen that:

- a **frequency table** provides a neat summary of the data and this leads to a representative histogram

- care needs to be taken when drawing a histogram – **frequency densities (frequency/interval width)** should be used if the intervals vary in width

- a **frequency polygon** also diagramatically represents the data and this needs to be plotted above the *mid-points* of intervals

- a **cumulative polygon** (plotted above the **ends** of intervals) enables us to find percentiles, lower quartiles, medians and upper quartiles

- a **box and whisker diagram** illustrates the range, the quartiles and the median

- a **stem and leaf diagram** is useful for illustrating discrete data.

ANSWERS

Practice questions A

1   (a)   This is a small sample, so 6 intervals will do:

$100 \le x < 300,$ $300 \le x < 500,$
$500 \le x < 700,$ $700 \le x < 900,$
$900 \le x < 1100,$ $1100 \le x < 1300.$

(b)   This is a large sample, so 11 intervals will probably be preferable:

$100 \le x < 200,$ $200 \le x < 300,$
......
$1100 \le x < 1200.$

2   Worm B is 1 cm $\le x < 2$ cm

Worms A and C in 2 cm $\le x < 3$ cm.

Practice questions B

1   *Relative frequency:* 0.1, 0.2, 0.38, 0.18, 0.06, 0.04, 0.04

Cumulative frequency: 5   15   34   43   46   48   50

2   *Frequency:* 3, 5, 11, 17, 4

Practice questions C

1 *Frequency:* 2, 6, 8, 10, 4, 4, 2   **2** 20
   Cumulative
   frequency: 2 8 16 26 30 34 36

Practice questions D

1 *Frequency density:* 4.8, 5.3, 4.4, 2.2, 1.9

2 5 + 15 + 20 + 10 + 50 = 100

3

Intervals	Width	Frequency	Frequency density
$14\frac{1}{2}$–$16\frac{1}{2}$	2	2	1
$16\frac{1}{2}$–$19\frac{1}{2}$	3	10	$3\frac{1}{3}$
$19\frac{1}{2}$–$25\frac{1}{2}$	6	6	1
$25\frac{1}{2}$–$35\frac{1}{2}$	10	7	0.7

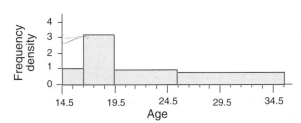

Practice questions E

1 15, 25, 35, 45, 55   **2** (a) 21 (b) 22.5 (c) 8.7 (d) 5.46

Practice questions F

1 *End points:*   7   9 11 13 15 17   **2** *End points:*   10 14 18 22 26
   Cumulative frequency: 5 22 45 59 68 70   *Cumulative frequency:*   9 26 49 63 66

Practice questions G

1 *End points:*   20   30   40   50   60   70
   Cumulative frequency: 5   13   30   42   47   50
   Percentage cumulative frequency: 10   26   60   84   94   100

   (a) ~ 26.3g   (b) ~ 29.4g   (c) ~ 37.1g   (d) ~ 46.3g

Practice questions H

1 (a)
0	5 7 9
1	1 1 1 3 5 7 8 8 9
2	0 1 1 2 2 5
3	1 9

2 (a) 30   (b) 15

   (b) • Students appear to watch more
         programmes than senior citizens
       • Students viewing habits more varied.

Practice questions I

1 (a)

w:	$2.5 \le w < 7.5$	$7.5 \le w < 12.5$	$12.5 \le w < 17.5$	$17.5 \le w < 22.5$	$22.5 \le w < 27.5$
f:	7	18	37	29	9
cf:	7	25	62	91	100
Ends:	7.5	12.5	17.5	22.5	27.5

(b) lower quartile ~12.5
median ~15.5
upper quartile ~19.5
(Remember to plot 7 against 7.5,
25 against 12.5 etc.)

(c)

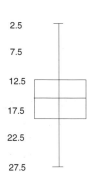

2 (a) (i) lower quartile ~16
(ii) median ~24
(iii) upper quartile ~34.

(Remember to plot 11 against 10, 35 against
20, etc.)

(b)

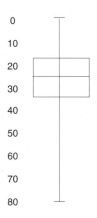

3 (a) 22

(b) 61

(c) (i) 13   (ii) 4

(d) • Range of males is less (1 to 43 against
5 to 61)

• Males peak between 10 and 19, whereas
females peak between 30 and 38.

• Females read more magazines on
average than males.

3

Representing data II: Measures of location

In the previous section we examined a variety of methods for presenting data diagrammatically. We saw that such diagrams help us to identify the important features of a data set: patterns and trends, for example. The next stage in this descriptive process is to extract detailed statistical information from a data set in order to quantify such patterns and trends, and to describe in numerical terms the key features of some set of data.

In this section we will be looking at the calculation of three types of average: the mean, the median and the mode. When the data are precisely known, we will see how to calculate exactly these three measures of location. However, when the data provided is imprecise, then we will have to *estimate* the mean and the median. As for the mode, we'll be able to go no further than the modal class.

Finally we will look at the effect of linear transformations on these three measures of location (i.e. coding).

The arithmetic mean
OCR **S1** 5.11.1 (d),(f)

One of the first calculations we would normally require is to determine a typical, or average, value. In statistics, there are three different measures of average and here we examine the first of these: the **arithmetic mean**.

The **arithmetic mean** (usually just referred to as 'the mean') is defined as follows:

For the set of n items of data $\{ x_1, x_2, x_3, \ldots x_n \}$

$$\bar{x} = \frac{x_1 + x_2 + x_3 + \ldots + x_n}{n}$$

which, using the sigma notation from P1, abbreviates to:

$$\text{Arithmetic mean } \bar{x} = \frac{1}{n} \sum_{i=1}^{n} x_i$$

where:

$x_1, x_2, \ldots$ are the values of the variate X

Σ is the summation symbol telling us to add all the items of data together

n is the symbol for the number of items in the data set, e.g. in the example that follows $n = 5$

$\bar{x}$ (pronounced 'x bar') is the standard symbol for the arithmetic mean of a set of sample data.

Example	Find the arithmetic mean of the following heights of males (cm)

{ 168.3, 175.2, 195.3, 163.0, 175.0 }

Solution	$\bar{x} = \frac{1}{5}$ (168.3 + 175.2 + 195.3 + 163.0 + 175.0)

$$\Rightarrow \bar{x} = \frac{1}{5} \times 876.8$$

$$\Rightarrow \bar{x} = 175.4 \text{ cm} \quad (1 \text{ d.p.})$$

The arithmetic mean for a *frequency distribution* can best be found through a simple example.

Example	Some students collect the information shown in Table 3.1 about the numbers of children in their families. Calculate the mean of these family sizes.

Table 3.1

Number of children in family (x)	Number of families f
1	3
2	9
3	5
4	2
5	1

Solution	To calculate the mean of these family sizes, we need to add up all the family sizes and divide by the number of families surveyed.

This can be written as:

$$\bar{x} = \frac{1 \times 3 + 2 \times 9 + 3 \times 5 + 4 \times 2 + 5 \times 1}{3 + 9 + 5 + 2 + 1}$$

or, symbolically, $\bar{x} = \dfrac{\sum fx}{\sum f} = \dfrac{49}{20} = 2.45$

This is often most conveniently worked out by using another column on the distribution table.

Table 3.2

Number of children in family (x)	Number of families f	fx
1	3	1 × 3 = 3
2	9	2 × 9 = 18
3	5	3 × 5 = 15
4	2	4 × 2 = 8
5	1	5 × 1 = 5
	$\sum f = 20$	$\sum fx = 49$

$$\therefore \bar{x} = \frac{\sum fx}{\sum f} = \frac{49}{20} = 2.45$$

Practice questions A

1 Find the mean of the following:

5, 3, 4, 4, 2, 6, 7, 1, 2, 6

2 Fifty students took a test and were awarded marks as follows:

Mark:	20	21	22	23	24	25
Frequency:	4	7	11	16	8	4

Find the mean.

3 Five children have a mean weight of 49 kg. When Sally joins the group, the mean weight drops to 48 kg. How much does Sally weigh?

Interpreting and using the mean   OCR **S1** 5.11.1 (b)

As with every statistic it is important not just to be able to get the right answer, but also to be able to use the statistic properly.

To illustrate the use of the mean let us return to our samples of males and females. If we perform the corresponding calculations (and you may want to do this yourself just to get practice), we find that:

$\bar{x}$ = 175.4 cm for the sample of males

$\bar{x}$ = 166.8 cm for the sample of females.

We could reasonably conclude that, on average, the mean height of males is almost 9 cm higher than that of females. This is not to say that all males are taller than females. We know, in fact, that this is not the case from the original data. We are saying that *typically* a male will be taller than a female.

The arithmetic mean for aggregated data   OCR **S1** 5.11.1 (d),(f)

In the calculations above we used the raw data to calculate the mean. Where we have data in the form of a frequency table the method of calculation is slightly different. The reason for this becomes apparent if we examine Table 3.3 which shows the frequency table for male heights.

Table 3.3	Frequency table for males
Heights cm	*Number of males*
155–	1
160–	3
165–	8
170–	12
175–	14
180–	9
185–	2
190–	0
195–	1
Total frequency	50

For example, that there are eight males with a height somewhere between 165 and 170 cm but we do not know from the table precisely what these heights are.

To get round the problem we make a simplifying assumption that all eight males had a height of 167.5 cm, i.e. the value of the mid-point of the internal. Similarly we calculate the midpoint values for the other intervals.

And so the calculation is as follows:

Table 3.4		Calculation of mean for grouped data (heights of adult males)	
	Number of males f	Mid-interval value x	Frequency × mid-interval value fx
155–	1	157.5	157.5
160–	3	162.5	487.5
165–	8	167.5	1340.0
170–	12	172.5	2070.0
175–	14	177.5	2485.0
180–	9	182.5	1642.5
185–	2	187.5	375.0
190–	0	192.5	0
195–	1	197.5	197.5
	$\Sigma f = 50$		$\Sigma fx = 8755.0$

$$\text{Mean} = \frac{\Sigma fx}{\Sigma f} = \frac{8755.0}{50} = 175.1 \text{ cm}$$

The mean based on the aggregated data gives a slightly different value from that based on the raw data because we used mid-points.

And so aggregated data only provide an *estimate* of the sample mean.

Practice questions B

1   Estimate the means for the following samples:

   (a) Body temperature (°C) of a group of pupils:

Temperature:	36.0–36.4	36.5–36.9	37.0–37.4	37.5–37.9
Frequency:	3	18	15	4

   (b) Weights of worms:

Weight (g):	$0 \le w < 4$	$4 \le w < 8$	$8 \le w < 12$	$12 \le w < 16$	$16 \le w < 20$	$20 \le w < 24$
Frequency:	8	11	32	28	15	6

2   Refer back to practice question H2 in the last section (page 21). Find the mean numbers of letters received. (Find the true mean – do not estimate it.)

The median

OCR **S1** 5.11.1 (d)

The median is the second type of average used in statistics.

Consider the data below, relating to the heights of nine adult males.

    168.3   167.2   169.1   167.7   166.5   169.1   195.3   163.0   165.0 cm

The mean is calculated as 170.1 cm but it is clear that this is not really a typical height, given that eight of the nine observations fall below the mean. It is evident on reflection that this has happened because one number, 195.3, is much higher than the others.

The median is an alternative measure of average that can be useful in such situations. *The median is literally the middle value in the data set: that is a value such that there is the same number of values above the median as below.* Unlike the mean the median always splits the ordered set of data into two equal parts.

The median is calculated by first ranking the data from lowest to highest.

163.0   165.0   166.5   167.2   167.7   168.3   169.1   169.1   195.3 cm

and then locating the middle item. In this case item number 5 is the middle item as there are the same number of observations below this value as above it. In this example there are four people below the median height and four people above it. (Note that we carefully distinguish between the **median item** and the **median value**.)

The only point to note relates to the determination of the median item. In general we can determine which observation this will be by using:

$$\frac{n+1}{2}$$

where n is the total number of observations in the data set.

In our example this is:

$$\frac{9+1}{2} = \frac{10}{2} = 5$$

So the observation in the middle of the data set is item 5. We hit a snag if we have an even number of items. Suppose that the data here also included someone of height 164.0 cm. Then we have:

163.0, 164.0, 165.0, 166.5, 167.2, 167.7, 168.3, 169.1, 169.1, 195.3

Counting in from the ends, we find that we cannot identify a middle item. But we can find a *middle pair* (here 167.2 and 167.7).

The convention is that we take as median the number half-way between the two (167.45) or, if you like, the *mean* of the middle pair: $\frac{167.2 + 167.7}{2} = 167.45$.

There are two points to note:

● The median here is not a member of the original data set.

● We have here 10 items; the rule $\frac{n+1}{2}$ would give us $\frac{11}{2} = 5.5$.

We still use that rule and take it to indicate that the middle pair is given by the items ranking either side of 5.5 – i.e. 5th and 6th – and we would determine the median value from these items.

Practice questions C

1   Find the median for the following sets of data:

(a) 3, 3, 4, 9, 16

(b) 1, 4, 5, 6, 6, 8

2   Find the median for the following sets of data. (Don't forget to *re-order* first of all.)

(a) 3, 5, 2, 9, 6, 4, 3, 3, 3, 4

(b) 100m times for Carl Lewis:
10.0, 10.1, 9.89, 10.01, 9.95.

Calculating the median from aggregated data $\boxed{\text{OCR S1 5.11.1 (d)}}$

As with the mean we may need to calculate the median from grouped data. Again, we must remember that in such a case the value we obtain will only be an estimate of the sample median value. Let us examine Table 3.5 which again shows the frequency distribution of heights of adult males.

Table 3.5

Heights cm	Number of males	Cumulative frequency
155–	1	1
160–	3	4
165–	8	12
170–	12	24
175–	14	38
180–	9	47
185–	2	49
190–	0	49
195–	1	50
Total	50	50

First we identify the median item. Here,

$$\frac{n+1}{2} = \frac{50+1}{2} = 25.5$$

As we are dealing with aggregated data we must identify the interval in which the median item falls. From the cumulative frequencies we see that the median item occurs in the interval $175 \le x < 180$ cm. We refer to this interval as the **median interval**.

We now know that the median item is one of the 14 values falling in this interval. Although there are 14 items in this interval we do not know what their exact values are. Again, as with the grouped mean, we make a simplifying assumption to enable us to perform the necessary calculations.

We assume that these 14 items are spread equally throughout the interval. Given that we have 14 items and that the interval covers 5 cm, this means that we can work out how far apart each item is from its neighbours.

$$\text{Distance apart} = \frac{\text{Interval width}}{\text{Interval frequency}} = \frac{5}{14} = 0.357 \text{ cm}$$

From Table 3.5 we know that the median item (25.5) is actually 1.5 items into the median interval (given that there were a total of 24 items up to this interval). Given that each item in this interval is 0.357 cm apart, this means that the median item is 0.536 cm into this interval (that is, 1.5×0.357).
Given that the interval starts at 175 cm this implies the median value is:

$$175 + 0.536 \text{ cm} = 175.54 \text{ (to 2 d.p.)}$$

The method here is called **linear interpolation**. The assumption being made in this method is that the data in an interval is spread evenly throughout that interval.

Estimating the median from the ogive $\boxed{\text{OCR S1 5.11.1 (d)}}$

As well as calculating a value for the median of an aggregated data set, it is also possible to estimate the median value from the appropriate **cumulative**

frequency curve, also known as an **ogive** (pronounced 'oh-jive'). The cumulative frequency curve is a modified version of the cumulative frequency polygon and simply involves smoothing the lines of the polygon into a curve.

Figure 3.1 shows the percentage ogive derived from the percentage frequency polygon in Fig. 2.6.

Figure 3.1

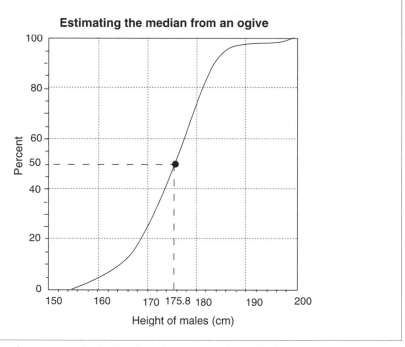

We know that the median is defined as the middle item. This actually corresponds to the middle quartile (at the 50% position on the percentage cumulative frequency axis). If we draw a line from this point to the ogive curve and down to the height axis we estimate the median to be roughly 175.8 cm.

We should remember that estimating the median from the ogive graph will be inaccurate because of the limitations of the scales on the graph.

It should also be noted that if we use the formula $\frac{n+1}{2}$ here to find which is the median item, we would get 50.5. The loss of accuracy by using 50 (the more 'obvious' choice from the diagram) is not significant and it is perfectly acceptable to use 50 as the middle value.

Practice questions D

1   For each of the following, find the median:

(a)
Goals scored	0	1	2	3	4
Frequency	6	4	5	2	1

(b)
Magazines per week	0	1	2	3
Number of people	8	3	2	2

(c)
Diameter of trees (cm)	$33 < d \leq 34$	$34 < d \leq 35$	$35 < d \leq 36$	$36 < d \leq 37$
Number of trees	6	35	38	21

2 80 snails are weighed.

Weight (g):	$0 \leq w < 20$	$20 \leq w < 40$	$40 \leq w < 60$	$60 \leq w < 80$	$80 \leq w < 199$
Frequency:	5	24	33	12	6

Draw a cumulative curve and hence estimate the median.

3 The number of fleas counted on 80 dogs is shown in the following table.

Number of fleas	0–2	3–5	6–8	9–11	12–14	15–17	18–20
Number of dogs	6	8	13	22	17	14	0

Estimate the median number of fleas per dog.

The mode

The third measure of average that we consider is the **mode**. The mode is defined as the value that occurs *most frequently* in a set of data. The mode is useful only when we are dealing with a discrete variable. For example, for the average number of exam passes at a school the mode would indicate the number of exams that most pupils passed. In this sense it can be regarded as a typical value.

One of the drawbacks of the mode is that a distribution may have two values sharing the highest frequency. Such a distribution is sometimes described as **bimodal** – having two modes. If there are more than two values sharing the highest frequency, it is usual to admit defeat and say that the distribution has no mode.

The modal class

For continuous data, grouped into a frequency table, although we cannot find a single value as the mode, it is possible to refer to the **modal class**, i.e. the class with the highest frequency, e.g. in Table 3.3, the modal class is 175– with a frequency of 14. Of course, like its discrete counterpart, the modal class may not be uniquely defined.

Practice questions E

1 Refer back to practice questions D1(a) and (b) on page 35. What are the modes?

2 Refer back to practice question D1(c) on page 35. What is the modal class?

3 The annual salaries of the workers in a factory were:

20 apprentices	£4500 each
15 semi-skilled workers	£9000 each
10 skilled workers	£13500 each
2 foremen	£15000 each
1 manager	£25000

(a) What is the modal salary?

(b) What is the median salary?

(c) What is the mean salary?

4 The number of insects on 50 dandelions were counted:

```
2 4 3 2 0 1 4 5 2 0
1 2 4 3 3 2 1 0 1 2
2 2 3 5 2 4 0 0 1 2
3 1 2 5 4 4 0 1 1 2
2 1 2 1 0 0 2 4 5 3
```

Copy and complete the following frequency table:

Number of insects:	0	1	2	3	4	5
Frequency:	8					

(a) What is the mean number of insects per plant?

(b) What is the median?

(c) What is the mode?

Comparing the different measures of average

Considering that there are three main measures of average, we must pose the question: which measure of average do you use? The answer, as always with statistics, is that it depends.

It depends both on the data that you are analysing and what the results of the analysis will be used for.

All three statistics have their particular uses and their disadvantages, and the only real answer is to consider carefully the problem you are working on.

- *The mean* is the most commonly used statistic, often chosen because it is the only one which uses the entire data set in its calculation. Consequently it can often be used in additional calculations. The problem with the mean, as we have seen, is that it is easily distorted by relatively few extreme values.

- *The median* is less affected by extremes and provides some information about the spread of values in the data set. The difficulty with the median is that it only uses part of the data.

- *The mode* tends to be useful in fairly limited circumstances, mostly when dealing with discrete data.

Use of coding and scaling in calculating $\bar{x}$

Look at the following example.

Example	Calculate the mean of the discrete data given in the following frequency table.

x	132–135	136–147	148–156	157–159
f	6	12	38	8

Solution

It should be clear that the mean is around 150. The method of **coding** assumes a sensible value for $\bar{x}$ and deducts this value from all the items of data, finds the average of the resulting data and adds it on at the end.

In table form:

x	f	midpoint of interval, m	$m - 150$	$(m - 150) \times f$
132–135	6	133.5	−16.5	−99
136–147	12	141.5	−8.5	−102
148–156	38	152	2	76
157–159	8	158	8	64
	64			−61

$$\text{Coded mean} \quad = \frac{-61}{64} = -0.95$$

$$\text{Correct mean} \quad = 150 - 0.95 = 149.05$$

It is possible to use a similar technique, called **scaling**, whereby you reduce all the items of data by a factor.

Example	Calculate the mean of:
	6200, 5100, 7800, 9300, 8200

Solution	We could proceed by dividing each item by 1000, finding the mean of the resulting numbers and then multiplying by 1000 at the end.

Scaled data is   { 6.2, 5.1, 7.8, 9.3, 8.2 }

$\bar{x}$ for scaled data is $\dfrac{36.6}{5} = 7.32$

therefore $\bar{x}$ for the original data is $7.32 \times 1000 = 7320$

Practice questions F

1 (a) Calculate the mean of the following data:

x	2	3	4	5	6
f	5	7	19	8	1

   (b) *Write down* the means of the following:

   (i)
y	12	13	14	15	16
f	5	7	19	8	1

   (ii)
g	20	30	40	50	60
f	5	7	19	8	1

   (iii)
h	7	9	11	13	15
f	5	7	19	8	1

2 Use a coding method to calculate the mean of the following:

x	1501	1503	1504	1507	1509
f	5	17	21	6	1

3 Consider the following data:

x	5	7	9	11	13	15	17
f	2	15	24	26	17	12	4

   (a) (i)   What is the mean of x?
   (ii) What is the median?
   (iii) What is the mode?

   (b) *Write down* the mean, mode and median for:

y	55	75	95	115	135	155	175
f	2	15	24	26	17	12	4

SUMMARY EXERCISE

1 Calculate the mean for females based on Table 2.2 from the last section (i.e. using the frequency table you found in Exercise 2 at the end of Section 2). Why are the means based on the grouped data slightly different from those based on the raw data?   **C 3.2**

2 Find the mean of the following:

   3, 8, 4, 9, 6, 8, 7, 7, 2

   What is the median value?

3 Find the mean and median of the following:

   4, 8, 11, 12, 6, 1

4 Find the mean of this data:

x	5	6	7	8	9	10
f	2	7	12	10	5	4

5 Estimate the mean and median of the following frequency distribution:

x	0–4	4–8	8–12	12–16	16–20
f	3	8	19	14	6

6 Use sensible coding to find $\bar{x}$ for each of the following:

   (a)
x	235–240	241–250	251–260	261–265
f	12	40	45	8

   (b) { 42 350, 36 890, 61 350, 53 400, 48 610 }

7 If $\bar{x}$ is the mean of
$$\{ x_1, x_2, x_3, \dots x_n \}$$
find in terms of $\bar{x}$ the means of:
(a) $\{ x_1 + a, x_2 + a, x_3 + a, \dots x_n + a \}$
(b) $\{ bx_1, bx_2, bx_3, \dots bx_n \}$

8

Population of village	$0 \le p < 50$	$50 \le p < 200$	$200 \le p < 1000$	$1000 \le p < 5000$
Number of villages	5	11	17	7

Estimate the median village size.

9 A number of cats are weighed:

30% weigh 2 kg or less
80% weigh 5 kg or less
Estimate the median cat weight.

10 Twenty items were weighed and gave:

Mean = 10 kg, median = 11 kg, mode = 2 kg.

A 21st item was found and added to the list. It weighed 9 kg. Which of the above three measures of location *must* change as a result?

11 Population of a number of towns:

Population	≤ 5000	$5000 < x \le 10000$	$10000 < x \le 50000$	$50000 < x \le 200000$
Frequency	73	256	1273	941

There are also 11 towns whose populations exceeds 200000. **C** 3.2

(a) Describe three difficulties you would have in drawing a histogram to represent the data.
(b) Estimate the median town size.

SUMMARY

In this section we have looked at three types of average – the mean, the median and the mode.

- the **mean** is the 'usual' type of average
- the **median** is the middle value of the ranked data
- the **mode** is the observation that occurs most frequently.

When the data are *precisely known:*

- we calculate the **exact value of the mean** by using the formula $\bar{x} = \dfrac{\sum fx}{\sum f}$

- we calculate the **exact value of the median** by finding the median item i.e. the $\left(\dfrac{n+1}{2} \right)$th item, where n is the number of observations in the data set

- we calculate the **exact value of the mode** (or modes) by writing down the number (or numbers) which occur the most frequently

When the data are *imprecisely known* (i.e. given in interval form):

- we calculate an **estimate of the mean** by using *midpoints* in the formula $\bar{x} = \dfrac{\sum fx}{\sum f}$

- we calculate an **estimate of the median** *either* by using linear interpolation *or* by drawing a cumulative curve and finding the 50th percentile

- the mode cannot be found – we can only give the modal class.

Finally, we have seen how a linear transformation affects the three measures of location. If the linear transformation is given by $x \rightarrow ax + b$, then each of the three measures of location (m) is given by $m \rightarrow am + b$.

ANSWERS

Practice questions A

1 4

2 22.58

3 43 kg

Practice questions B

1 (a) 36.95

(b) 11.96

2 14.8

Practice questions C

1 (a) 4

(b) 5.5

2 (a) 3.5

(b) 10.0

Practice questions D

1 (a) 1

(b) 0

(c) ~35.2

2 ~47 (Plot 5 against 20, 29 against 40, etc.)

3 ~10.3

Practice questions E

1 0 and 0

2 $35 < d \le 36$

3 (a) £4500

(b) £9000

(c) £8645.83

4 8, 10, 15, 6, 7, 4

Mean = 2.12, median = 2, mode = 2

Practice questions F

1 (a) 3.825

(b) (i)   13.825 ($y = x + 10$)

(ii) 38.25 ($g = 10x$)

(iii) 10.65 ($h = 2x + 3$)

2 $1500 + \dfrac{191}{50} = 1503.82$

3 (a) (i)   10.86

(ii) 11

(iii) 11

(b) 113.6, 115, 115 ($y = 10x + 5$)

Representing data III: Measures of dispersion

INTRODUCTION In Section 3 we examined the three statistical measures of average and saw that such averages do not always reflect a 'typical' value in the data set. The individual items in a data set may vary considerably from the average. We need to introduce further statistics which allow us to describe and quantify such variation, or dispersion as it is known in statistics.

There are three measures of dispersion that we'll consider here – the range, the interquartile range and the standard deviation. The range is the easiest to calculate, but the interquartile range requires either the use of a cumulative curve or linear interpolation. As for the standard deviation, that either requires a table of long-hand calculations or an ability to feed the data into a calculator and let it do the work for you!

That being done we'll consider the types of skewness shown by any distribution and the effect that any rogue readings (outliers) might have on any of the measures of dispersion.

The range

OCR **S1** 5.11.1 (a),(b)

The range is the simplest measure of dispersion. For a set of data it is defined as follows:

> Range = largest item – smallest item

It is very easily calculated but tells us nothing about the *distribution* of data within the set of data. For the data sets in Section 2, the ranges are:

Range of male heights = 195.3 – 159.8 = 35.5 cm

Range of female heights = 181.9 – 153.3 = 28.6 cm

This suggests that male heights are more variable than female heights.
In some circumstances the range can give a rough and ready measure of the 'spreadout-ness' of data but would not be of such use where data is very skewed.

Practice questions A

1 Find the range of the following sets of data:
   (a) 5, 8, 12, 14, 21
   (b) 3, 24, 19, 36, 5, 0

x	$5 \leq x < 10$	$10 \leq x < 15$	$15 \leq x < 20$	$20 \leq x < 25$
f	5	8	11	3

(c)

The standard deviation

OCR **S1** 5.11.1 (d),(f)

The standard deviation is the measure of variation most often used in statistics: it measures variation around the arithmetic mean.

Standard deviation is defined by the following formula:

$$\text{Standard deviation} \quad s = \sqrt{\frac{\sum_{i=1}^{n}(x_i - \bar{x})^2}{n}}$$

where the set of data is: $\{x_1, x_2, x_3 \dots x_n\}$ (i.e. n items of data)

From the formula above, if we square both sides we get:

$$\text{Variance } s^2 = \frac{\Sigma(x_i - \bar{x})^2}{n}$$

This quantity is called the **variance** of the data. In later work, the variance will turn out to be used more often than the standard deviation as a measure of disperson.

An example showing how to calculate s and how to set out the calculation now follows.

Example

Calculate the standard deviation of the set of data { 4, 5, 7, 9, 10 }.

Solution

The calculations are set out in Table 4.1 which is followed by an explanation.

Table 4.1	Calculation of standard deviation	
x_i	$x_i - \bar{x}$	$(x_i - \bar{x})^2$
4	$4 - 7 = -3$	9
5	$5 - 7 = -2$	4
7	$7 - 7 = 0$	0
9	$9 - 7 = 2$	4
10	$10 - 7 = 3$	9
$\Sigma x_i = 35$		$\Sigma(x_i - \bar{x})^2 = 26$

$$\Rightarrow \bar{x} = \frac{35}{5} = 7 \Rightarrow s^2 = \frac{26}{5} = 5.2$$

$$\Rightarrow s = 2.28 \text{ (2 d.p.)}$$

In the first column are the individual items of data. The first column is totalled and the result divided by 5 to give $\bar{x}$, the mean. This is then used in the second column to calculate the deviation of each item of data from the mean.

If these deviations are totalled, the result will always be zero (or thereabouts, if rounding has been necessary), i.e. the positive deviations cancel the negative deviations. For this reason it is necessary to square the deviations and this is done in column 3.

Column 3 is totalled and then, beneath the table, the final calculation of s is carried out. The total is divided by 5, the number of items of data (this is averaging the squared deviations) and the result of this is then square-rooted (to offset the original effect of squaring).

This method of calculating the standard deviation looks straightforward enough with the easy numbers in this example. But, if the mean is not a whole number, subtracting it from every value and then squaring the result involves a lot of decimal places and/or a lot of rounding. There is an easier method!

The calculating formula

OCR **S1** 5.11.1 (d),(f)

It can be shown that the formula for the variance s^2 is also given by:

$$s^2 = \frac{\sum_{i=1}^{n} x_i^2}{n} - \overline{x}^2$$

which in practice is a much simpler formula to use.

Example

Calculate s for this set of data, first using the definition of s and then using the calculating formula:

{168.3, 175.2, 195.3, 163.0, 175.0}

Solution

The calculations are as follows.

Table 4.2 **Calculation of standard deviation**

x_i	$x_i - \overline{x}$	$(x_i - \overline{x})^2$	x_i^2
168.3	−7.06	49.84	28 324.89
175.2	−0.16	0.03	30 695.04
195.3	+19.94	397.60	38 142.09
163.0	−12.36	152.77	26 569.00
175.0	−0.36	0.13	30 625.00
$\Sigma x_i = 876.8$		$\Sigma(x_i - \overline{x})^2 = 600.37$	$\Sigma(x_i^2) = 154\ 356.02$

Both methods use the first column to find $\overline{x}$ in the same way:

$$\overline{x} = \frac{876.8}{5} = 175.36$$

The first method, using the definition of s, then uses this value of $\overline{x}$ to work out the values of $(x_i - \overline{x})$ and $(x_i - \overline{x})^2$ in columns 2 and 3.

Then the total of column 3 has to be divided by the number of items:

$$s^2 = \frac{600.37}{5} = 120.074 \quad \Rightarrow s = 10.96$$

The second method, using the calculating formula, doesn't need columns 2 and 3 – just column 4 (which is much more straightforward to work out).

It then continues:

$$s^2 = \frac{154\ 356.02}{5} - 175.36^2 = 120.074 \implies s = 10.96$$

In a problem with more values in it than in this one, the saving of effort over the first method is very welcome. It should be your usual way to work out a standard deviation.

Practice questions B

1 Find the mean, variance and standard deviation of the following sets of data:
   (a) 5, 3, 4, 4, 2, 6, 7, 1, 2, 6
   (b) 28, 27, 26, 32, 25, 26, 20, 21, 19, 22, 23, 24, 27, 26, 31, 30
   (c) 16 yrs 8 mths, 16 yrs 4 mths, 17 yrs, 17 yrs 2 mths, 15 yrs 11 mths, 15 yrs 7 mths, 16 yrs 3 mths, 17 yrs.

Calculating the standard deviation for aggregated data

OCR **S1** 5.11.1 (d),(f)

Example

Let us return once more to the frequency table for the heights of adult males where we have already calculated the mean (from the frequency table) as 175.1 cm. The method we adopt is similar in approach to that of calculating the mean from a frequency table: we use midpoint values to estimate the individual data items.

Just as we had a formula for calculating the standard deviation for raw data so we have a formula for the aggregated data:

$$\text{Standard deviation} = \sqrt{\frac{\Sigma f x^2}{\Sigma f} - \left(\frac{\Sigma f x}{\Sigma f}\right)^2} = s$$

Table 4.3 shows the various calculations that we need to undertake to work out the standard deviation from the frequency table.

Table 4.3 Calculation of the standard deviation for grouped data (heights of males)

Interval (cm)	Frequency f	Midpoint x	fx	fx^2*
155–	1	157.5	157.5	24,806.25
160–	3	162.5	487.5	79,218.75
165–	8	167.5	1340.0	224,450.00
170–	12	172.5	2070.0	357,075.00
175–	14	177.5	2485.0	441,087.50
180–	9	182.5	1642.5	299,756.25
185–	2	187.5	375.0	70,312.50
190–	0	192.5	0	0
195–	1	197.5	197.5	39,006.25
Total	50		8755.0	1,535,712.50

*You can calculate fx^2 as $x \times fx$.

We are now in a position to put all the appropriate values into the formula and to work out the answer.

For the adult males:

$$\text{Standard deviation} = \sqrt{\frac{\Sigma fx^2}{\Sigma f} - \left(\frac{\Sigma fx}{\Sigma f}\right)^2} = \sqrt{\frac{1,535,712.5}{50} - \left(\frac{8755}{50}\right)^2}$$

$$= \sqrt{30,714.25 - 30,660.01} = \sqrt{54.24} = 7.36 \text{ cm} = s$$

That is, the standard deviation of heights of adult males is 7.36 cm.

Practice questions C

1 Find the mean, variance and standard deviation of the following data:

(a)

Number of cars per min	40	41	42	43	44
Frequency	17	32	38	31	12

(b) Body temperature (°C) of a group of pupils.

Temperature	36.0–36.4	36.5–36.9	37.0–37.4	37.5–37.9
Frequency	3	18	15	4

(c)

Length (cm)	$8 \le l < 10$	$10 \le l < 12$	$12 \le l < 14$	$14 \le l < 16$	$16 \le l < 18$	$18 \le l < 20$
Frequency	8	11	32	28	15	6

As with the mean and median we should see the calculation of the standard deviation for grouped data as an estimate of the value which would be obtained directly from the raw data. Manual calculation of the standard deviation for the raw data, however, is extremely time-consuming and tedious for anything more than a few numbers and the calculation based on the grouped data is usually used.

Some problems involve combining two sets of data and finding means and variances.

Example

(a) The lengths of 20 rods are measured in cm

The measurements are summarised by $\Sigma x = 285$, $\Sigma x^2 = 4250$.
Calculate $\bar{x}$ and s^2.

(b) The sample is enlarged by a further 10 rods.

The lengths of these 10 rods may be summarised by $\Sigma x = 155$, $\Sigma x^2 = 2300$. Calculate the mean and variance of the whole sample of 30 rods.

Solution

(a) $\bar{x} = \dfrac{285}{20} = 14.25$ cm

$s^2 = \dfrac{\Sigma x^2}{n} - \bar{x}^2 = \dfrac{4250}{20} - 14.25^2 = 9.4375$ cm

(b) $\sum x$ for the 30 rods is found by adding the $\sum x$ figures for both parts of the sample. $\sum x = 285 + 155 = 440$

$$\Rightarrow \bar{x} = \frac{440}{30} = 14.67 \text{ (2 d.p.)}$$

Similarly, adding the two sets of $\sum x^2$ figures:

$$\sum x^2 = 4250 + 2300 = 6550 \quad \Rightarrow s^2 = \frac{6550}{30} - 14.67^2 = 3.22.$$

Practice questions D

1   Find the mean, variance and standard deviation for the following sets of data:

(a) Sample of 10 $\sum x = 48$ $\sum x^2 = 268$

(b) Sample of 20 $\sum x = 11.4$ $\sum x^2 = 7.64$

(c) 20 mice are weighed in grammes and give:

   $\sum x = 464$, $\sum x^2 = 10790$

(d) Marks achieved by 15 pupils in a test:

   $\sum x = 118$, $\sum x^2 = 2236$.

2   A sample of 30 readings gives $\sum x = 182$, $\sum x^2 = 1276$.

A further sample of 10 readings gives $\sum x = 39$, $\sum x^2 = 185$.

Find the mean, variance and standard deviation for the 40 readings.

Using your calculator to find mean and standard deviation

OCR **S1** 5.11.1 (f)

It is instructive to be able to calculate the mean and standard deviation long-hand but, in practice, you let your calculator do the work for you.
With most calculators, the procedure is as follows:

1   Put the standard deviation mode on.

   (This is usually indicated by SD.)

2   Clear the memory.

   (Usually this is either $\boxed{\text{inv}}$ $\boxed{\text{AC}}$ or $\boxed{\text{Shift}}$ $\boxed{\text{AC}}$.)

3   Feed in your figures.

   (If the sample is *small* this usually requires $\boxed{x}$ $\boxed{\text{M+}}$

   If the sample is *large* this usually requires $\boxed{x}$ $\boxed{\times}$ $\boxed{f}$ $\boxed{\text{M+}}$

4   Key out the values of n (useful for a check), $\bar{x}$ and s (written $x\sigma_n$ on your calculator).

   (Usually this involves either:

   $\boxed{\text{inv}}$ $\boxed{}$, $\boxed{\text{inv}}$ $\boxed{}$ and $\boxed{\text{inv}}$ $\boxed{}$ respectively, or
   $\quad\quad n \quad\quad\quad\quad \bar{x} \quad\quad\quad\quad\quad\quad x\sigma_n$

   $\boxed{\text{K out}}$ $\boxed{3}$, $\boxed{\text{shift}}$ $\boxed{1}$ and $\boxed{\text{shift}}$ $\boxed{2}$ respectively.)
   $\quad\quad n \quad\quad\quad\quad \bar{x} \quad\quad\quad\quad\quad\quad x\sigma_n$

   It is very important that you can use your calculator properly, so a set of examples on which you can practise now follows.

Examples on mean, variance and standard deviation

Work through the following and check that you agree with the given answers.

1   3,   4,   5,   6,   7,   8,   9

Ans: $n = 7$, $\bar{x} = 6$, $s = 2$, Variance = 4

2   4,   5,   8,   3,   7,   6,   6,   7,   1,   4

Ans: $n = 10$, $\bar{x} = 5.1$, $s = 2.022$ ∴ Variance = 4.09

3   1,   0,   2,   0,   3,   3,   0,   0,   1,   4

Ans: $n = 10$, $\bar{x} = 1.4$, $s = 1.428$ ∴ Variance = 2.04

Remember to feed in the zeros as well.

4

x	2	4	6	8	10
f	3	5	11	4	2

Ans: $n = 25$, $\bar{x} = 5.76$, $s = 2.141$ ∴ Variance = 4.58 (to 2 d.p.)

You begin by feeding in:

| 2 | × | 3 | M+ |

You must *not* begin by feeding in:

| 6 | M+ |

5

x	3	6	9	12	15
f	8	11	17	9	5

Ans: $n = 50$, $\bar{x} = 8.52$, $s = 3.568$, Variance = 12.73 (to 2 d.p.)

6   Use your calculator to check the mean and standard deviation for males, as given in Table 4.3.

Practice questions E

1   *Use your calculator* to work out the mean, variance and standard deviation for the following sets of data:

(a)

Number of blooms	0	1	2	3	4	5	6	7
Frequency	3	5	7	14	38	42	23	2

(b)

x	8	9	10	12	14
f	3	14	11	7	5

(c) Sample of 14 $\sum x = 84$ $\sum x^2 = 612$

(d) Sample of 10 $\sum x = 39$ $\sum x^2 = 185$

(e) Sample of 10 $\sum x = 48$ $\sum x^2 = 268$

(f) Sample of 30 $\sum x = 182$ $\sum x^2 = 1276$

(g) *Time (mins):* $0 \le t < 2$ $2 \le t < 4$ $4 \le t < 6$ $6 \le t < 10$ $10 \le t < 20$
    Frequency: 17   23   42   8   10

(h) 30 spiders weighed in g: $\sum x = 240$, $\sum x^2 = 5998$

(i)	Time (sec)	$0 \leq t < 0.2$	$0.2 \leq t < 0.4$	$0.4 \leq t < 0.6$	$0.6 \leq t < 1.0$	$1.0 \leq t < 2.0$
	Frequency	7	18	36	8	1

2 (a) 2, 3, 5, 7, 13.   What are the mean and standard deviation?

   (b) 3, 4, 6, 8, 14.   What are the mean and standard deviation?

   (c) What will be the mean and standard deviation of 4, 5, 7, 9, 15?

3 Fifty items are measured and give a mean of 2.7 cm and a standard deviation of 1.3 cm.
Later it was found that all 50 measurements were underestimated by 0.2 cm.
What will be the mean and standard deviation now? (See question 2?)

The interquartile range

OCR **S1** 5.11.1 (b),(d)

So far in looking at dispersion we have concentrated on measuring dispersion around the mean. We shall now examine how dispersion around the median can be quantified using a statistic known as the **interquartile range.**

The interquartile range is defined as:

> Interquartile range = $Q_3 - Q_1$
>
> where Q_3 represents the upper (75%) quartile
>
> and Q_1 represents the lower (25%) quartile

The upper and lower quartiles can be calculated in any one of four ways:

Method 1: From the raw data we can use the same method that we used for the median. Instead of counting along the ordered data looking for the item in the middle (the 50% item) we would look instead for the 25% item for the lower quartile and 75% item for the upper quartile.

Method 2: From the frequency table we can use a formula similar to the one we used for the median. You will remember the formula we used to determine the median item was:

$$\text{median item} = \frac{(n + 1)}{2}$$

For the lower quartile we would want item $\dfrac{(n + 1)}{4}$ instead (because we now want the item which is $\frac{1}{4}$ of the ordered data set).

For the upper quartile we require item $\dfrac{3(n + 1)}{4}$ because we now seek the item which is $\frac{3}{4}$ of the ordered data set.

Method 3: From the percentage ogive; again, instead of finding the median at the 50% point we would find the lower quartile at the 25% point and the upper quartile from the 75% point.

Method 4: Use of linear interpolation as explained when calculating the median, where for grouped data we assume that the data is evenly spread through that interval.

Which method we use will depend on the problem we are looking at and the form of any analysis and calculations that we have already undertaken.

Table 4.4 shows the quartiles for males based on Table 2.5 in Section 2, using Method 2.

Table 4.4	Quartiles for males (all figures in cm)

Median = Q_2 = 175.5
Lower quartile = Q_1 = 170.3
Upper quartile = Q_3 = 180.1
∴ Interquartile range = $Q_3 - Q_1$ = 9.8

The table shows three quartiles (remember that the median is effectively the middle quartile) as well as the interquartile range.

The corresponding results for female heights are in Table 4.5.

Table 4.5	Quartiles for females (all figures in cm)

Median = Q_2 = 167.5
Lower quartile = Q_1 = 161.9
Upper quartile = Q_3 = 172.1
∴ Interquartile range = $Q_3 - Q_1$ = 10.2

Let us look and see what the interquartile range can tell us about the data.

The interquartile range is the difference between the upper and lower quartile. It represents the central 50% of the distribution, that is half of the data set falls into the interquartile range. So, as a measure of dispersion the interquartile range always contains the same proportion of the data set, the middle 50% around the median.

For males the interquartile range is 9.8, whilst for females it is 10.2. The fact that the interquartile range is smaller in males indicates that the items in the middle of the data set are relatively closely clustered around the median. There is less of a difference between the top 25% height figure and the bottom 25% height figure than in females.

Practice questions F

1 Find the interquartile range for the following sets of data:

(a) 3, 5, 6, 9, 10, 10, 12, 14, 15

(b) 4, 6, 7, 9, 12, 14

(c)

x	2	3	4	5	6	7	8
f	5	8	11	17	13	4	2

(d)

x	4	5	6	7	8
f	1	7	11	4	3

2 Draw a cumulative curve for the following data:

Time (sec):	$0 < t \le 20$	$20 < t \le 40$	$40 < t \le 60$	$60 < t \le 80$	$80 < t \le 120$
Frequency	8	25	33	59	35

Hence estimate: (a) lower quartile (b) upper quartile (c) interquartile range.

3 Time (mins) for 80 football fans to get through turnstiles:

Time (mins)	1–4	5–8	9–12	13–16	17–20	21–30
Number people	4	6	16	30	17	7

Draw a cumulative curve for the above data.

Hence estimate: (a) lower quartile (b) upper quartile (c) interquartile range.

4 Weight of 40 people:

Weight (kg)	$45 \le w < 55$	$55 \le w < 65$	$65 \le w < 75$	$75 \le w < 85$	$85 \le w < 95$
Number of people	0	6	15	14	5

Use linear interpolation to estimate:

(a) the mean

(b) the lower quartile

(c) the upper quartile

(d) the interquartile range.

5 Use linear interpolation to estimate the interquartile range:

x	$0 \le x < 2$	$2 \le x < 4$	$4 \le x < 6$	$6 \le x < 8$	$8 \le x < 10$	$10 \le x < 12$
f	5	8	17	11	6	3

6 In a survey conducted by a football club, 100 supporters were asked to record the length of time it had taken them to queue and get through the turnstiles. The frequency table below gives their times, in minutes, rounded to the nearest minute.

Time taken	1 to 5	6 to 10	11 to 15	16 to 20	21 to 25	26 to 30	31 to 35
Number of people	5	7	18	35	19	12	4

(a) State clearly the boundaries of the modal class of this distribution.

(b) Estimate the mean time taken.

(c) Complete the cumulative frequency table below:

Time taken (less than)	5.5	10.5	15.5	20.5	25.5	30.5	35.5
Number of people	5						

(d) Draw a cumulative graph to represent the data.

(e) Use your graph to estimate:

   (i)   the median time

   (ii) the interquartile range

   (iii) the number of spectators who had to queue for 25 minutes or longer.

(f) The football club estimates that, with an identity card system for entrance to football matches, it will take, on average, 60% longer for spectators to gain entrance to its ground. Using your answer to (b), estimate the mean time it will take spectators to get into the club's ground once the identity system has been installed.

Skewness

We have seen over the past few sections how we can use the statistics we have introduced to describe features of the data set. One feature that we often require relates to the general shape of the distribution.

In the examples we have used in earlier sections, the shape of the distribution of the data can vary considerably.

As we have seen, the mean and median for a data set are often different. The difference will be largely due to extreme values at one end of the distribution. Suppose we had a few extremely high values in the data set. When working out the mean these numbers will be included in the calculation (remember that the mean uses all the data). These extremely high values, however, will not be included in the calculation for the median because in its calculation we only count along to the middle of the data.

So, a few extremely high values will tend to pull the mean higher than the median. Extremely low values, on the other hand, will pull the mean below the median. We can imagine encountering three general types of distribution which are illustrated in Figs 4.1 to 4.3.

Figure 4.1

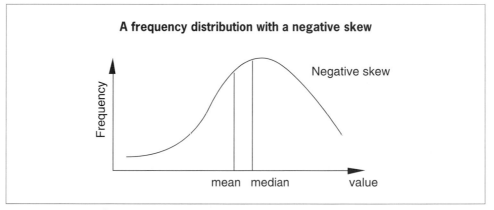

Figure 4.1 illustrates a relatively small number of low values pulling the mean below the median. The mean will take a lower value than the median and we refer to this type of distribution as having a **negative skew**. The more extreme the lower values, the more the distribution will concentrate on the left-hand side and the larger the negative skewness will become.

Figure 4.2

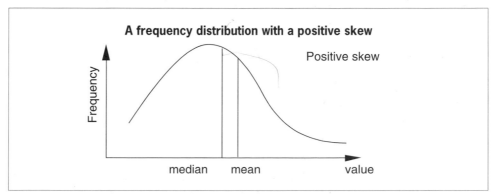

In Figure 4.2 there is a relatively small number of extremely high values, which will pull the mean above the median, giving a **positive skew**. The distribution in such a case will be concentrated on the left-hand side.

Figure 4.3

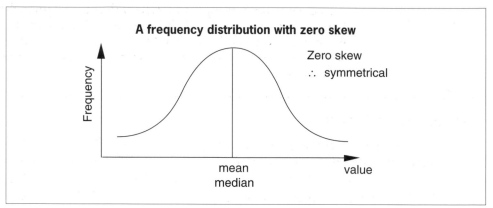

A symmetrical distribution, as in Figure 4.2, will have a zero skew, with the mean and median equal (or at least approximately so).

Practice questions G

1   For each of the following sets of data find
    (or estimate) the mean and the median.
    In each case describe the skewness implied.

(a) 2, 2, 3, 5, 14

(b)

x	3	5	7	9	11
f	12	10	6	5	4

(c)

x	0–2	2–4	4–6	6–8	8–10
f	3	5	12	5	3

(d)

x	5–9	9–13	13–17	17–21	21–25
f	3	7	14	5	1

Outliers

We saw briefly in Section 2 how outliners (or rogue points) can upset a distribution. Let's look at this in a little more detail.

Suppose we asked 13 of our friends how many letters they received last week and their replies were:

   0, 0, 1, 3, 3, 5, 6, 6, 7, 7, 8, 24 and 29

Clearly the last two figures are out of line with the remainder (probably these two friends had a birthday during the week). So let's call them *rogue points* (or *outliers*) and see what we have without them. For the remaining set of figures 0, 0, 1, 3, 3, 5, 6, 6, 7, 7 and 8 we get median = 5, lower quartile = 1, upper quartile = 7 and mean = $4\frac{2}{11}$.

This gives the box and whisker plot shown in Figure 4.4.

Figure 4.4

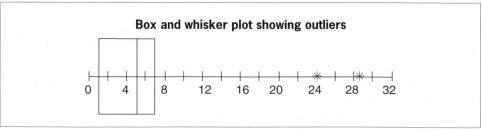

(where the two outliers are denoted by *).

This is a much fairer representation of the set of figures than would have been the case had we considered all 13 readings. For a start we would have got a mean of 9, which is above all but two of the readings! It would also have suggested a positive skew whereas the 'reasonable' plot above suggests a negative skew.

Practice questions H

1   A group of 9 senior citizens were asked how many cups of tea they drank last week. Their replies were: 2, 5, 40, 41, 41, 43, 43, 44 and 45.

   (a) Find their mean consumption. Comment on the reliability of the mean as a fair average.

   (b) Identify two outliers. What is the mean now? Comment.

   (c) Excluding these two outliers find the (i) median (ii) lower quartile and (iii) upper quartile.

   (d) Draw the corresponding box and whisker plot and highlight the two outliers.

2   The number of fleas on 12 dogs was as follows: 0, 0, 1, 3, 3, 3, 4, 4, 4, 4, 31, 56.

   Identify two outliers and, excluding them, find the mean and standard deviation for the number of fleas on a dog.

SUMMARY EXERCISE

1   Calculate the standard deviation of these data sets:

   (i)   using the definition of standard deviation

   (ii) using the calculating formula

   (a) { 2, 7, 12, 13, 21 }

   (b)

x	4	5	6	7	8
f	3	8	12	10	2

2   Calculate the standard deviation for the heights of females using the frequency data produced in Exercise 1 on p. 24.

3   Calculate the mean and variance of the data in this frequency distribution:

x	5–	10–	15–	20–	25–	30–
f	4	9	15	13	7	3

4   The weights (in kilograms) of two groups of students are measured and summarised:

	n	$\sum x$	$\sum x^2$
Group A	10	550	32 500
Group B	15	970	63 500

   (a) Calculate the mean and variance of each group's weights.

   (b) The two groups are now treated as a single sample of 25 students. Calculate its mean and variance.

5   Give **one** advantage and **one** disadvantage of grouping data into a frequency table.

   The table shows the trunk diameters, in centimetres, of a random sample of 200 larch trees.

Diameter (cm)	15–	20–	25–	30–	35–	40–50
Frequency	22	42	70	38	16	12

   Plot a cumulative frequency curve of these data.

   By use of this curve, or otherwise, estimate the median and the interquartile range of the trunk diameters of larch trees.

   A random sample of 200 spruce trees yields the following information concerning their trunk diameters, in centimetres.

Minimum	Lower quartile	Median	Upper quartile	Maximum
13	27	32	35	42

   Use this data summary to draw a second cumulative frequency curve on your graph.

 C 3.2

   Comment on any similarities or differences between the trunk diameters of larch and spruce trees.

   [AEB 1993]

6 In an investigation of delays at a roadworks, the times spent, by a sample of commuters, waiting to pass through the roadworks were recorded to the nearest minute. Shown below is part of a cumulative frequency table resulting from the investigation.

Upper class boundary	2.5	4.5	7.5	8.5	9.5	10.5	12.5	15.5	20.5
Cumulative number of commuters	0	6	21	48	97	149	178	191	200

(a) For how many of the commuters was the time recorded as 11 minutes or 12 minutes?

(b) Estimate:

    (i) the lower quartile,

    (ii) the 81st percentile, of these waiting times.

7 A railway enthusiast simulates train journeys and records the number of minutes, x, to the nearest minute, trains are late according to the schedule being used. A random sample of 50 journeys gave the following times.

17	5	3	10	4	3	10	5	2	14
3	14	5	5	21	9	22	36	14	34
22	4	23	6	8	15	41	23	13	7
6	13	33	8	5	34	26	17	8	43
24	14	23	4	19	5	23	13	12	10

(a) Construct a stem and leaf diagram to represent these data.

(b) Comment on the shape of the distribution produced by your diagram. **C** 3.2

(c) Given that $\sum x = 738$ and $\sum x^2 = 16\,526$, calculate (to 2 decimal places) estimates of the mean and the variance of the population from which this sample was drawn.

(d) Explain briefly the effect that grouping of these data would have had on your calculations in (c).

8 (a) Explain how the median and quartiles of a distribution can be used when describing the shape of a distribution.

Summarised below is the distribution of masses of new potatoes, in grams to the nearest gram.

Mass (g)	Frequency
19 or less	2
20–29	14
30–39	21
40–44	34
45–49	39
50–59	42
60–69	13
70–79	9
80–89	4
90 or more	2

(b) Use interpolation to estimate the median and quartiles of this distribution. Hence describe its skewness.

(c) Draw a box and whisker plot to illustrate these data.

9

x	0–4	4–8	8–12	12–20
f	5	11	36	8

(a) Illustrate the above distribution with a histogram. What is the modal class?

(b) Draw a cumulative frequency polygon and hence estimate:

    (i) the median and

    (ii) the semi-interquartile range.

(c) Use your calculator to estimate the mean and standard deviation of x.

10 Estimate the interquartile range from the following data:

x	2–4	5–7	8–10	11–13
f	3	7	12	2

SUMMARY

In this section we have looked at three measures of dispersion – the range, the interquartile range and the standard deviation.

- The **range** is the largest item minus the smallest item.

- The **interquartile range** is $Q_3 - Q_1$, where the lower quartile Q_1 is item $\dfrac{n+1}{4}$

  and the upper quartile Q_3 is item $3\left(\dfrac{n+1}{4}\right)$.

- The interquartile range is probably best calculated from a cumulative curve but linear interpolation is a good alternative method.

- The **standard deviation** (s), which measures dispersion around the mean, is given by either:

$$s = \sqrt{\frac{\sum(x_i - \bar{x})^2}{n}} \qquad \text{or} \qquad s = \sqrt{\frac{\sum x_i^2}{n} - \bar{x}^2} \qquad \left(\text{where } \bar{x} = \frac{\sum x_i}{n}\right)$$

or (if frequencies are involved) by either:

$$s = \sqrt{\frac{\sum f(x_i - \bar{x})^2}{n}} \qquad \text{or} \qquad s = \sqrt{\frac{\sum f x_i^2}{n} - \bar{x}^2} \qquad \left(\text{where } \bar{x} = \frac{\sum f x_i}{n} \text{ and } n = \sum f_i\right)$$

- **Variance** = (standard deviation)^2 so standard deviation = $\sqrt{\text{Variance}}$

- **Using a calculator** to find the standard deviation usually requires the following steps:

  SD mode on
  Shift AC
  Key in data by x × f M+
  Key out 3 to check that n is correct
  shift 1   gives $\bar{x}$
  shift 2   gives $x\sigma_n$ (the calculator notation for s, the standard deviation)

- When n, $\sum x$ and $\sum x^2$ are given, **the calculator is usually used as follows**:

  SD mode on
  Shift AC
  n key in 3
  $\sum x$ key in 2
  $\sum x^2$ key in 1
  Then shift 1 for $\bar{x}$ and shift 2 for $x\sigma_n$ (i.e. s).

We have also seen that:

- a distribution is **negative skew** if mean < median

- a distribution is **positive skew** if mean > median

- **outliers** (or rogue points) can have an undue influence on a distribution and so are often best rejected.

ANSWERS

Practice questions A

1 (a) $21 - 5 = 16$

(b) $36 - 0 = 36$

(c) $25 - 5 = 20$

Practice questions B

1 (a) 4, 3.6, 1.897

(b) 25.4375, 13.621, 3.691

(c) 16 years 5.875 months,
40.36 months, 6.35 months

Practice questions C

1 (a) 41.915, 1.370, 1.17

(b) 36.95, 0.15, 0.387

(c) 13.98, 6.44, 2.538

Practice questions D

1 (a) 4.8, 3.76, 1.939 (3 d.p.)

(b) 0.57, 0.0571, 0.239 (3 d.p.)

(c) 23.2, 1.26, 1.122g (3 d.p.)

(d) $7\frac{13}{15}$, $87\frac{41}{225}$, 9.337 (3 d.p.)

2 5.525, 5.999, 2.449 (3 d.p.)

Practice questions E

1 (a) 4.291, 2.057, 1.434

(b) 10.35, 3.23, 1.80

(c) 6, $7\frac{5}{7}$, 2.777 [14 $\boxed{\text{K in}}$ n, 84 $\boxed{\text{K in}}$ Σx,
612 $\boxed{\text{K in}}$ Σx^2, etc]

(d) 3.9, 3.29, 1.814

(e) 4.8, 3.76, 1.939

(f) 6.067, 5.73, 2.393

(g) 5.1, 14.35, 3.788

(h) 8, $135\frac{14}{15}$, 11.659

(i) 0.457, 0.049, 0.2214

2 (a) 6, 3.899

(b) 7 (up one), 3.899 (same)

(c) 8 (up another one), 3.899 (same)

3 2.9 cm (up 0.2) and 1.3 cm (same)

Practice questions F

1 (a) $2\frac{1}{2}$th item $= 5\frac{1}{2}$

$7\frac{1}{2}$th item $= 13$

∴ Interquartile range $= 13 - 5\frac{1}{2} = 7\frac{1}{2}$

(b) $1\frac{3}{4}$th item $= 5\frac{1}{2}$

$5\frac{1}{4}$th item $= 12\frac{1}{2}$

∴ Interquartile range $= 12\frac{1}{2} - 5\frac{1}{2} = 7$

(c) $15\frac{1}{4}$th item $= 4$

$45\frac{3}{4}$th item $= 6$

∴ Interquartile range $= 6 - 4 = 2$

(d) $7 - 5 = 2$

2 (a) $\cong 44$

(b) $\cong 78$

(c) $\cong 34$

3 (Plot 4 above $4\frac{1}{2}$, 10 above $8\frac{1}{2}$, etc.)

(a) $\cong 11$

(b) $\cong 17$

(c) $\cong 6$

4 (a) 74.5

(b) $\cong 68$

(c) $\cong 81$

(d) $\cong 13$

5 $12\frac{3}{4}$th $= 2 + \dfrac{7\frac{3}{4}}{8} \times 2 = 3.9375$

$38\frac{1}{4}$th $= 6 + \dfrac{8\frac{1}{4}}{11} \times 2 = 7.5$

∴ Interquartile range $= 3.5625$

6 (a) $15.5 - 20.5$

(b) 18.4

(c) 5, 12, 30, 65, 84, 96, 100

(e) (i) $\cong 18.4$

(ii) $\cong 23 - 14.5 = 8.5$

(iii) $\cong 18$

(f) 29.44

Practice questions G

1   (a)   5.2, 3. Positive skew
     (b)   5.86 (2 d.p.), 5. Positive skew
     (c)   5, 5. Symmetrical
     (d)   14.2, 14.57 (2 d.p.). Slight negative skew

Practice questions H

1   (a) $33\frac{7}{9}$. This is unreliable, as seven figures are
        above this figure and only two below.

     (b)   2 and 5 are outliers. Mean becomes = $42\frac{3}{7}$,
        a much fairer average

     (c)   Median = 43, lower quartile = 41,
        upper quartile = 44

     (d)   (* = outliers)

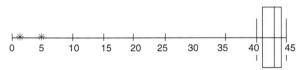

2   31 and 56 are outliers.

    Mean = 2.6,

    standard deviation = $\sqrt{2.44}$ = 1.562 (3 d.p.)

5
Probability

INTRODUCTION Whether we are explicitly aware of the fact or not, chance and uncertainty play an important part in our lives. When preparing for an exam we weigh up the chances of a particular topic appearing on the paper. When going to school or college to work each day we assess the chance that it might rain and we will need an umbrella. Such decisions have to be taken under conditions of uncertainty and it is frequently necessary to assess the likelihood of specific events occurring in the future. The analysis and quantifying of such assessments is known as probability theory, and forms the basis for important topics in statistics.

In this section we'll be meeting the idea of a sample space and an event. This will involve set notation so, should you have not met this before, you should first refer to the brief summary given in Appendix 1 at the end of the book. Then we shall see how to use tree diagrams and permutations and combinations when solving probability questions.

The idea of conditional probability may well be new to you but, essentially, it involves working out probabilities in the light of additional given information. For example, what is the chance of throwing 3 heads with 3 pennies? ($\frac{1}{8}$). But if I tell you you have thrown at least 2 heads, what's the chance you've thrown 3 heads now? ($\frac{1}{4}$).These types of probability questions require special techniques and we will be looking at them in detail.

Finally we will link conditional probability with the set notations introduced earlier and, in particular, investigate the ideas of mutually exclusive and independent events.

Trials, sample space and event

OCR **S4** 5.14.1

To illustrate the rules and language of probability we will consider some simple examples. An example which will be used throughout this section is as follows:

Eight cards numbered 1, 2, 3, ... , 8 but otherwise identical are placed in a container from which they can be selected randomly. In this way each card is *equally likely* to be selected.

A **trial** in probability theory is an action with several possible outcomes and for the set of 8 cards above a simple trial would be to select a single card from the container and observe the number written on it.

The set of possible outcomes for a trial is called the **sample space** (sometimes also the **possibility space**) and in this particular example it would be the set S given by:

$$S = \{\ 1,\ 2,\ 3,\ 4,\ 5,\ 6,\ 7,\ 8\ \}$$

Note that provided the selection is made randomly, each of the elements of this set is equally likely to occur. A sample space could therefore be defined as 'the set of possible outcomes of a trial'.

Any subset of a sample space is called an *event*

e.g. { 2, 4, 6, 8 }, or ∅ (the empty set)

(A summary of results and notation concerning sets is given in Appendix 1 at the end of the book for students who have not met them before.)

With our example of 8 numbered cards it is possible to devise more complicated trials which in turn will give rise to more complicated sample spaces and events.

Example	A card is selected, its number noted and then replaced. A second card is then selected and its number noted.

Describe the sample space and give an event associated with this sample space.

Solution	The sample space will consist of all ordered pairs of numbers (x, y) where they are both integers from 1 to 8 or more concisely

$$S = \{ (x, y) : \ 1 \le x \le 8, \ 1 \le y \le 8, \ x, y \text{ integers} \}$$

An example of an event associated with this S is 'the first number selected is less than the second number selected'. This can be described more formally by

$$A = \{ (x, y) : \ 1 \le x \le 8, \ 1 \le y \le 8, \ x, y \text{ integers}, \ x < y \}$$

Events and probabilities

OCR **S4** 5.14.1 (a)

Returning to the sample space $S = \{ 1, 2, \dots 8 \}$ for the single selection of a card, some events we could describe are:

A	$= \{ 2, 3, 5, 7 \}$	(a prime number is chosen)
B	$= \{ 1, 2, 3 \}$	(a card less than 4 is selected)
$\varnothing$		(no card is selected)
S		(a card is selected)

> If a sample space S consists of equally likely outcomes, then
> the probability P(X) of an event X is simply defined by the rule: $P(X) = \dfrac{n(X)}{n(S)}$

And so, for the events above:

$$\text{P}(A) \quad = \frac{n(\{ 2, 3, 5, 7 \})}{n(\{ 1, 2, 3, \dots 8 \})} \quad = \frac{4}{8} = \frac{1}{2}$$

$$\text{P}(B) \quad = \frac{n(\{ 1, 2, 3 \})}{n(\{ 1, 2, 3, \dots 8 \})} \quad = \frac{3}{8}$$

$$\text{P}(\varnothing) \quad = \frac{n(\varnothing)}{n(S)} \quad = \frac{0}{8} = 0$$

$$\text{P}(S) \quad = \frac{n(S)}{n(S)} \quad = 1$$

It should be clear from this that:

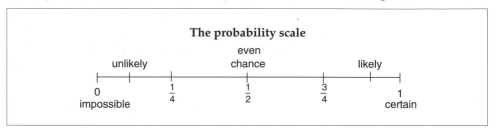

for any event X $0 \le P(X) \le 1$

An event with probability 0 is called *impossible* and an event with probability 1 is called *certain*. All other events will have a probability ranging from unlikely, through even chance, to likely, along a scale from 0 to 1 (see Fig. 5.1).

Figure 5.1

The probability scale

unlikely   even chance   likely

0 impossible $\frac{1}{4}$ $\frac{1}{2}$ $\frac{3}{4}$ 1 certain

Example

From the eight numbered cards, two cards are taken one after the other and where the first is replaced before the second is taken (this method of selection is called **sampling with replacement**). The numbers on the cards chosen are added together. Find the probability of obtaining a total score between 6 and 13.

Solution

A sample space of ordered pairs can be usefully represented as a set of points on a coordinate grid as in Fig. 5.2.

Figure 5.2

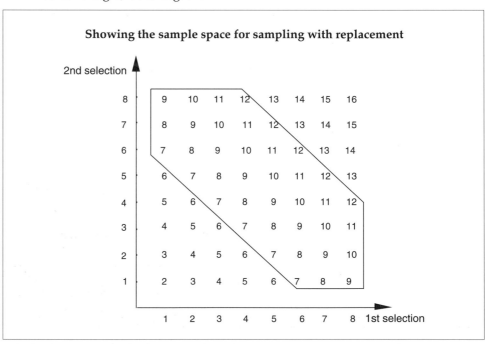

Showing the sample space for sampling with replacement

The ordered pairs which give a total satisfying $6 < \text{score} < 13$ are shown enclosed in the diagram and it can now be seen that:

$$P(6 < \text{score} < 13) = \frac{39}{64}$$

Compound events

Events can be combined using the same rules and notation as for sets, to form compound events.

Example	Let S $\quad = \{1, 2, 3, \ldots, 8\}$ and A $\quad = \{1, 3, 5, 7\}$ $\quad B$ $\quad = \{1, 2, 3\}$ $\quad C$ $\quad = \{7, 8\}$ Write down the events $A \cap B$, $A \cup C$, $B \cap C$, A', C', $A \cup C'$, $(A \cup C)'$ and find $P(A)$, $P(B)$, $P(C')$ $P(B \cap C)$, $P(A \cup B)$, $P(A \cap B)$. Calculate $P(A) + P(B) - P(A \cap B)$.

Solution	$A \cap B$ $\quad$ = elements in common to A and B = $\{1, 3\}$ $A \cup C$ $\quad$ = elements in A or B or both = $\{1, 3, 5, 7, 8\}$ $B \cap C$ $\quad$ = $\varnothing$ A' $\quad$ = elements which are in S but not in A = $\{2, 4, 6, 8\}$, $\quad\quad\quad$ called the **complement** of A C' $\quad$ = $\{1, 2, 3, 4, 5, 6\}$ $A \cup C'$ $\quad$ = $\{1, 3, 5, 7\} \cup \{1, 2, 3, 4, 5, 6\}$ = $\{1, 2, 3, 4, 5, 6, 7\}$ $(A \cup C)'$ = $(\{1, 3, 5, 7\} \cup \{7, 8,\})' = (\{1, 3, 5, 7, 8\})' = \{2, 4, 6\}$ $P(A)$ $\quad = \dfrac{4}{8} = \dfrac{1}{2}$ $P(B)$ $\quad = \dfrac{3}{8}$ $P(C')$ $\quad = \dfrac{6}{8} = \dfrac{3}{4}$ $P(B \cap C) = \dfrac{0}{8} = 0$ $P(A \cup B) = \dfrac{n(\{1, 3, 5, 7, 8\})}{n(S)} \quad = \dfrac{5}{8}$ $P(A \cap B) = \dfrac{2}{8} = \dfrac{1}{4}$ $P(A) + P(B) - P(A \cap B) = \dfrac{1}{2} + \dfrac{3}{8} - \dfrac{1}{4} = \dfrac{5}{8}$, which is the same as $P(A \cup B)$

Some important rules are suggested by this example.

Rule 1	For any event X from sample space S $P(X') = 1 - P(X)$

Rule 2	For any events X and Y from a sample space S $P(X \cup Y) = P(X) + P(Y) - P(X \cap Y)$

The truth of Rule 1 should be clear.

The truth of Rule 2 is more difficult to illustrate, but consider Fig. 5.3.

Figure 5.3

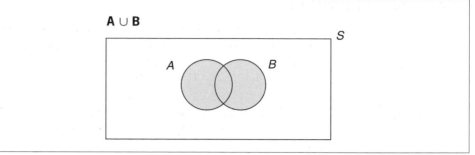

To evaluate P($A \cup B$) we could add up the probability in A and the probability in B. However in doing this, we would have added the intersection twice, so it therefore needs to be subtracted once.

In the previous Example we saw that $B \cap C = \varnothing$, and it followed that P($B \cap C$) = 0. Events which have this property are called *mutually exclusive*. Mutually exclusive events have the property that if one of them occurs then the other cannot possible occur. As an example, if you choose a card from an ordinary pack of playing cards, the events A = 'picking a black card' and B = 'picking a red card' are mutually exclusive.

Practice questions A

1 A fair dice has three faces marked 1, one face marked 2, and two faces marked 4.

Another fair dice has four faces marked 1 and two faces marked 2.

These two dice are thrown and the *total* score recorded.

- Set up a sample space for the possible totals.
- Hence find the probability that the total score is even.

2 The illustrated spinner is spun twice and the *total* score is recorded.

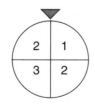

- Set up a sample space for the possible totals.
- What is the probability that the total exceeds 3?

3 (a) If P(A) = 0.6, find P(A').

(b) If P(A) = 0.6, P(B) = 0.3 and P(A ∩ B) = 0.5, find P(A ∪ B).

(c) If P(A) = 0.6, P(C) = 0.2 and P(A ∪ C) = 0.8, find P(A ∩ C).

What can be said about sets A and C?

4 $P(A) = \frac{1}{3}$ $P(B) = \frac{1}{2}$ $P(A \cup B) = \frac{5}{6}$

Prove that A and B are mutually exclusive and find P(A').

5 The sample space S is given by

S = {0, 1, 2, 3, 4, 5, 6, 7, 8, 9}

and subsets A, B and C are given by

A = {0, 1, 2, 3}, B = {2, 3, 4, 5, 6} and C = {6, 7, 8}.

Find:

(a) A'   (b) A' ∩ B   (c) B ∪ C

(d) (B ∪ C)'   (e) A ∩ B'.

Hence find:

(f) P(A')   (g) P(A' ∩ B)   (h) P(B ∪ C)

(i) P[(B ∪ C)']   (j) P(A ∩ B').

Using tree diagrams

Tree diagrams are useful in situations where there is a sequence of trials although their use becomes unwieldy if the sequence is very long or there are more than a few options at each trial. However the method is useful and illustrates several important ideas.

Example

Two cards are drawn from the eight numbered cards with replacement (i.e. the first is replaced before the second is drawn).

Find the probability that:

(a)   both cards have numbers greater than 5

(b)   at least one of the numbers is a 6.

Solution

(a)   We are only concerned here with the card being > 5 or ≤ 5, so there are two options at each trial.

The tree diagram (the visual aid to solving the problem) is as follows:

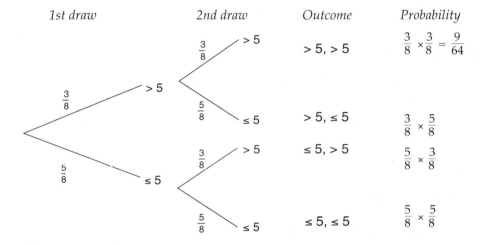

The complete tree diagram shows the outcomes at the ends of the branches and their probabilities on the branches. So at the first trial the outcome is > 5 (i.e. 6, 7 or 8) with probability $\frac{3}{8}$ as shown. After the first trial has taken place, the first card is replaced and so the probabilities at the second trial are going to be the same.

However, this time we have a second trial for each of the possible outcomes of the first trial – i.e. each branch now branches into two. The various outcomes and their probabilities are then listed at the ends of the final branches as shown.

The final column illustrates an important point which pervades probability theory.

If we consider the outcome '> 5, > 5' this is a shorthand for '> 5 *and* > 5' and in the final column '*and*' has been replaced by '$\times$'. Whenever '*and*' occurs in probability problems, it will be necessary to multiply the probabilities together.

We will return to this point later when the idea of independence is met but for the moment the following rule will be sufficient.

> **'and' ⇔ ×**

The answer to the question 'What is the probability that both cards are greater than 5?' is now seen to be $\frac{9}{64}$.

Before leaving the example it should be noted that the probabilities in the final column add up to 1 and this should always be the case for a correct tree diagram.

(b)   In this part we are only concerned with whether a number is a 6 or not, so the tree diagram has the form:

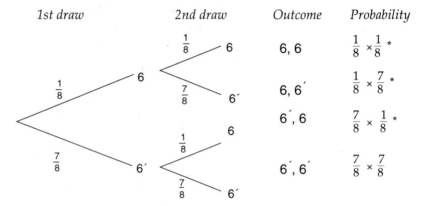

To find the probability that at least one of the numbers is a 6, we need to consider the ways in which this can happen.

The first three rows marked with * are all cases where one (or more) of the cards is a 6 and provide alternative ways of satisfying the condition.

In other words, 'at least one card is a six' corresponds to:
6, 6 *or* 6, 6′ *or* 6′, 6

and P(at least one card is a 6)
$$= \frac{1}{8} \times \frac{1}{8} + \frac{1}{8} \times \frac{7}{8} + \frac{7}{8} \times \frac{1}{8} = \frac{15}{64}$$

This example illustrates the important rule that:

> **'or' ⇔ +**

i.e. whenever 'or' occurs in a probability problem it corresponds to *adding* probabilities.

It should be noted at this point that an alternative way of solving this problem is to use the complement of the event 'at least one card is a six', i.e. 'no card is a six'.

$$\text{P(no card is a 6)} \quad = \text{P(6′ and 6′)} = \frac{7}{8} \times \frac{7}{8} = \frac{49}{64}$$

Hence   P(at least one card is a 6) $= 1 - \frac{49}{64} = \frac{15}{64}$, as before.

This method of using the complement (and not forgetting to subtract from 1 at the end!) is particularly useful in problems involving the phrase 'at least'.

Example

Two cards are drawn randomly from an ordinary pack of playing cards but this time a card is not replaced before the next is removed (*sampling without replacement*). Find the probability that:

(a) The two cards are of the same colour

(b) Just one of the cards is a spade.

Solution

(a) *1st draw*   *2nd draw*   *Outcome*   *Probability*

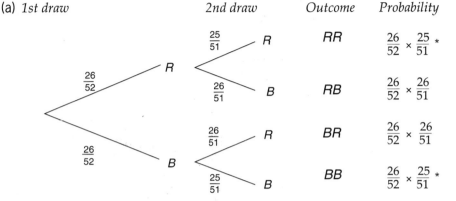

	RR	$\frac{26}{52} \times \frac{25}{51}$ *
	RB	$\frac{26}{52} \times \frac{26}{51}$
	BR	$\frac{26}{52} \times \frac{26}{51}$
	BB	$\frac{26}{52} \times \frac{25}{51}$ *

You should note carefully how the fact of 'not replacing' the first card affects the probabilities at the second draw.

The outcomes marked * correspond to two cards being of the same colour and we have:

$$\text{P(cards are same colour)} = \frac{26}{52} \times \frac{25}{51} + \frac{26}{52} \times \frac{25}{51} = 2 \times \frac{26}{52} \times \frac{25}{51} = \frac{25}{51}$$

(b) The relevant tree diagram is:

1st draw   *2nd draw*   *Outcome*   *Probability*

1st draw	2nd draw	Outcome	Probability
$\frac{13}{52}$ S	$\frac{12}{51}$ S	SS	$\frac{13}{52} \times \frac{12}{51}$
	$\frac{39}{51}$ $\overline{S}$	$S\overline{S}$	$\frac{13}{52} \times \frac{39}{51}$ *
$\frac{39}{52}$ $\overline{S}$	$\frac{13}{51}$ S	$\overline{S}S$	$\frac{39}{52} \times \frac{13}{51}$ *
	$\frac{38}{51}$ $\overline{S}$	$\overline{S}\,\overline{S}$	$\frac{39}{52} \times \frac{38}{51}$

The relevant outcomes are marked *, and so:

$$\text{P(just one card is a spade)} = \frac{13}{52} \times \frac{39}{51} + \frac{39}{52} \times \frac{13}{51} = 2 \times \frac{13}{52} \times \frac{13}{17} = \frac{13}{34}$$

Practice questions B

1 Anabel has a $\frac{3}{4}$ probability of passing her driving test. Bert has a $\frac{3}{5}$ probability of passing his test.

   - Illustrate this situation with a tree diagram.
   - What is the probability that:
   (a) both pass   (b) neither pass
   (c) Anabel passes but Bert fails?

2 A bag contains 6 red and 8 green counters. Two are chosen at random without replacement.

   Illustrate the situation with a tree diagram.

   What is the probability that:

   (a) both are green   (b) just one is green?

3 A barrel contains 20 numbered discs 1, 2, 3, … 20. Two discs are taken out 'blind' and if you have a pair of prime numbers you WIN! Find the probability of winning.

4 A fairground stall consists of a barrel full of a large amount of 2p, 5p and 10p coins in the ratio 3:4:5. You are 'dealt' two coins at random. If their product is odd, then you win. Show that the probability of winning is a meagre $\frac{1}{9}$.

5 I stand a $\frac{3}{4}$ chance of passing my driving test. If I fail, I stand a $\frac{9}{10}$ chance of passing next time. Whatever happens, I am not going to take it more than twice.

   By drawing a tree diagram, find the probability that I pass.

6 Two buses are due at 12.00. The probability I get on the first one is $\frac{2}{3}$. If that one is full, the probability I get on the second one is $\frac{3}{4}$. Otherwise I have had it! What's the probability I get on a bus?

Permutations and combinations

OCR **S1** 5.11.2 (a),(b),(c),(d)

These can also be used when solving certain types of probability problems.

A **permutation** of some objects is essentially an **arrangement** of them. If we consider say 5 objects then we could arrange them in a row in 5! (= 120) ways. When you see 5!, read 'five factorial'.

Since we could fill the first position in the row in 5 ways, then we would have 4 choices for the second position and so on giving $5 \times 4 \times 3 \times 2 \times 1 = 5!$ ways in all.

Consider now 5 objects, but only three spaces. The first space could be filled in 5 ways, the second in 4 ways and the third in 3 ways giving $5 \times 4 \times 3$ ways altogether.

The number of permutations (or arrangements) of 3 objects from 5 objects is therefore $5 \times 4 \times 3$ which can be written as

$$\frac{5 \times 4 \times 3 \times 2 \times 1}{2 \times 1} \qquad \text{or} \qquad \frac{5!}{2!}$$

Similarly for 6 objects and 4 spaces this reasoning would give us $\frac{6!}{2!}$.

In general the number of permutations ^nP_r of r **distinguishable** objects out of n objects is:

$$^nP_r = \frac{n!}{(n-r)!}$$

Indistinguishable objects

It may be that some objects are **indistinguishable**. For example, how many permutations are there of the letters *ABBC* where it is not possible to distinguish between the two *B*'s?

To answer this question, pretend for a moment that we can distinguish the *B*'s by labelling them B_1, B_2. Then the arrangements would be 4! in number

i.e. $A \ B_1 \ B_2 \ C$, $A \ B_1 \ C \ B_2$, etc.

but the list would also include $A \ B_2 \ B_1 \ C$, $A \ B_2 \ C \ B_1$ and so on. If the *B*'s were not distinguished, these would be the same arrangements, namely $A \ B \ B \ C$ and $A \ B \ C \ B$. We have an answer which is two times too large.

The correct number of permutations is $\dfrac{4!}{2!}$.

If we had 9 objects where 4 were indistinguishable, we would similarly get $\dfrac{9!}{4!}$.

We could have objects $A \ B \ B \ B \ C \ D \ D \ D \ D$.

The result would be $\dfrac{9!}{3! \ 4!}$ where the factor 3! corresponds to the *B*'s and 4! corresponds to the *D*'s.

In general for *n* objects, *r* of which are the same, we have $\dfrac{n!}{r!}$ permutations and as can be seen above the result extends.

Example

Find the number of arrangements of the letters of the word:

S T A T I S T I C S

Solution

We have 10 letters and this includes:

3 S's, 3 T's, 1 A, 1 C, 2 I's

giving $\dfrac{10!}{3! \ 3! \ 2!} = 50 \ 400$

Example

Find the number of ways in which 11 boys can stand in a row if two of them refuse to stand next to each other.

Solution

This problem is difficult to solve directly.

The following approach is useful when a restriction is involved. Consider the number of arrangements if the boys are stuck together as a single entity. Call them Andy and Bill.

(*AB*) (remaining 9 boys)

How many arrangements?

There are now 10 objects altogether giving 10! arrangements.

However for each of these we could also have the equivalent but with *A* and *B* in reverse order, i.e.

(*BA*) (remaining 9 boys)

Hence in all, there are $2 \times 10!$ ways of arranging the boys so that Andy and Bill are next to each other. The total number of arrangements without restriction is $11!$

The difference between them is those arrangements where the boys are separated.

Hence number of arrangements $= 11! - (2 \times 10!) = 10! \,(11 - 2)$

$$= 9 \times 10! = 32\ 659\ 200$$

A *combination* of some objects is equivalent to a **selection**. As an example consider the following:

Out of a collection of 5 novels by my favourite author, I would like to choose 3 to take on holiday. How many choices do I have?

In this case it is not too difficult to list the selections provided we are systematic.

Call the books $A \;\; B \;\; C \;\; D \;\; E$

then the choices are:

A B C	A C D	B C D	C D E
A B D	A C E	B C E	
A B E	A D E	B D E	

i.e. 10 in all.

Now if order had been important, for each of these selections or combinations there would be 6 arrangements or permutations of each.

So to arrive at the number of combinations we need to reduce the number of permutations by the factor $6 \; (= 3!)$

Hence $^5C_3$ the number of combinations of 3 out of 5 is given by

$$^5C_3 \;=\; \frac{^5P_3}{3!}$$

$$=\; \frac{5!}{2!\,3!} \;\; \left(\text{using the fact that } ^5P_3 \;=\; \frac{5!}{2!} \right)$$

$$=\; 10$$

Generalising for a selection ^nC_r of r things out of n distinguishable things we have:

$$^nC_r \;=\; \frac{n!}{(n-r)!\,r!}$$

Note: an alternative and commonly used notation for ^nC_r is $\binom{n}{r}$.

Example	3 people have to be chosen from a group of 8 as a committee. How many selections are possible.
Solution	$^8C_3 \;=\; \dfrac{8!}{5!\,3!} \;=\; 56$

Example	In the National Lottery, 6 numbers have to be selected from 49 numbers to scoop the jackpot.
	(a) How many possible selections are there?
	(b) If one ticket is bought, what is the probability of winning?

Solution	(a) There are $^{49}C_6$ possible selections:

$$^{49}C_6 = \frac{49!}{43!\,6!} = 13{,}983{,}816$$

(b) The probability of winning with one entry is $\dfrac{1}{13\,983\,816}$ or approximately 1 in 14 million.

A slightly more difficult problem arises if we are selecting from a set of objects which are not all different.

Example	How many selections of 3 letters are there from the word:

P R O B A B L E

Solution	Consider the types of selection as:

2 B's   + 1 other
1 B   + 2 others
0 B's   + 3 others

since it is the double B which causes the problem.

2B's and 1 other is 6 ways
(since the extra one is chosen from P R O A L E)
1B   and 2 others is $^6C_2$ ways
(since we now have to choose 2 from P R O A L E)
0 B's   and 3 others is $^6C_3$
(since now we choose 3 from P R O A L E)

Total number of ways is $6 + {}^6C_2 + {}^6C_3$

$$= 6 + 15 + 20 = 41$$

Example	Find the probability that a hand of 13 cards dealt from an ordinary pack will contain at least 11 spades.

Solution	We need to work out the probabilities of:

11 spades
12 spades
13 spades

and then add these.

$$P(\text{11 spades}) = \frac{^{13}C_{11} \times {}^{39}C_2}{^{52}C_{13}}$$

since we have to select 11 out of 13 spades and 2 out of 39 other cards and this out of a total of $^{52}C_{13}$ possible selections of 13 out of 52.

$$P(\text{12 spades}) = \frac{^{13}C_{12} \times {}^{39}C_1}{^{52}C_{13}}$$

$$P(\text{13 spades}) = \frac{^{13}C_{13} \times {}^{39}C_0}{^{52}C_{13}}$$

Hence P (11 or more spades)

$$= \frac{(^{13}C_{11} \times {}^{39}C_2) + (^{13}C_{12} \times {}^{39}C_1) + (^{13}C_{13} \times {}^{39}C_0)}{^{52}C_{13}}$$

$$= 9.2 \times 10^{-8} \quad \text{(highly unlikely to happen)}$$

Practice questions C

1 Evaluate:

(a) $6!$ (b) $\frac{18!}{3!15!}$ (c) $\frac{86!}{84!}$ (d) $\binom{7}{3}$ (e) $\binom{9}{4}$

2 How many different arrangements are there of the following:

(a) RIDDLE (b) NECESSARY?

3 Football pools – the 'Treble Chance'. How many 'lines' are required in the following 'perms':

(a) any 8 from 10
(b) any 8 from 12?

At 1p a line how much will it cost to perm any 8 from 15?

4 A committee of 3 girls and 2 boys has to be chosen from a group of 8 girls and 4 boys. How many ways can this be done?

5 A box contains 14 sweets. In how many ways can I choose a handful of 3 sweets?

6 There are 14 possible players for the First XI. In how many ways can the team be chosen?

If 4 of the possible players are 'certain', in how many ways can the remainder be chosen?

7 A committee of 5 girls and 3 boys has to be chosen from a group of 11 girls and 7 boys. How many ways can this be done?

If I 'fix it' so that 3 particular girls are bound to be on the committee, in how many ways can the remaining members of the committee be chosen?

8 There are 3 parrots, 4 cockatiels and 2 canaries in a pet shop. I choose three birds at random. What is the chance they are:

(a) all cockatiels
(b) 1 parrot and 2 other birds
(c) 2 canaries and 1 other bird
(d) all the same type of bird
(e) non parrots?

9 A society with 20 members contains 3 'smokers'. A committee of 4 is chosen. What is the chance that the committee contains exactly 2 smokers?

10 Hen pigeons occur with probability $\frac{1}{2}$. I place three pigeons in a box. What is the chance of a mating pair?

How many pigeons must I place in a box to be 99% sure of at least one mating pair?

Conditional probability and independence   OCR **S1** 5.11.2 (f), **S4** 5.14.1 (c)

A card is selected from an ordinary pack of cards. What is the probability that it is a king? The answer is clearly $\frac{4}{52} = \frac{1}{13}$.

Suppose however there is some additional information available about the card selected. Suppose, for example, that it is known to be a court card (i.e. a king, queen or jack). What is the probability of it being a king if it is known to be a court card?

The extra information increases the probability to $\frac{4}{12} = \frac{1}{3}$ since there are only now 12 possibilities for the card. In other words the sample space has been reduced to a set containing only 12 members.

Suppose instead it was known to be a black card. How would this affect the probability of it being a king? The sample space is now reduced to 26 cards, but the number of possible kings has been reduced to 2 since it is known that the chosen card is black.

$$\text{Hence P(king given black card)} = \frac{2}{26} = \frac{1}{13}$$

This is the same as the original result where no extra information was given. Sometimes extra information doesn't make a difference to the final probability.

The solution where extra information is provided in a problem is referred to as a *conditional* probability and such problems are often written in the form 'what is the probability of some event *given* some extra information'.

We write $P(A \mid B)$ for the probability of event A, given that B is known to have occurred and this is calculated from the formula:

$$\text{The probability of A given B} = P(A \mid B) = \frac{P(A \cap B)}{P(B)}$$

The Venn diagram, Fig. 5.4, provides a justification for this formula.

Figure 5.4

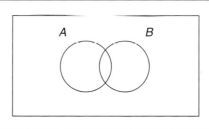

B is given (has occurred) so that we know we must be within that circle. What is the chance that A has happened given that we are in circle B?
It will equal the proportion of the circle B taken up by A

i.e. $\dfrac{P(A \cap B)}{P(B)}$

Effectively, B becomes the new sample space.

Example

From the eight numbered cards a card is selected. What is the probability it is higher than 4 given that it is even?

Solution

This is a fairly simple problem to solve and can probably most easily be done by counting methods. However the solution presented will use the formula as it illustrates a useful technique.

Let A be the event: 'The card is higher than 4.'

Let B be the event: 'The card is even.'

Then $P(A \mid B)$ is what is required, i.e. $\dfrac{P(A \cap B)}{P(B)}$

Now $P(B) = \left(\dfrac{n\,(\{2, 4, 6, 8\})}{n\,(\{1, 2, 3, 4, 5, 6, 7, 8\})} \right) = \dfrac{1}{2}$

Also $A \cap B$ is the event 'higher than 4 *and* even' and is satisfied by 6 and 8 only.

So $P(A \cap B) = \dfrac{2}{8}$

$\Rightarrow P(A \mid B) = \dfrac{2}{8} \div \dfrac{1}{2} = \dfrac{1}{2}$.

Example

Two cards are drawn from an ordinary pack without replacement. Find the probability that the first card is a spade given that the second card is not a spade.

Solution

Let $\quad A$ = '1st card is a spade'

Let $\quad B$ = '2nd card is not a spade'

then $\quad P(A \mid B) = \dfrac{P(A \cap B)}{P(B)}$ as usual.

$A \cap B$ is the event '1st card is a spade and 2nd card is not a spade' and can be abbreviated SS' using the notation introduced earlier.

$$P(SS') = \dfrac{13}{52} \times \dfrac{39}{51} = \dfrac{13}{68} = P(A \cap B)$$

$P(B)$ is obtained by considering the ways in which the second card can turn out to be a spade and a tree diagram helps to find that:

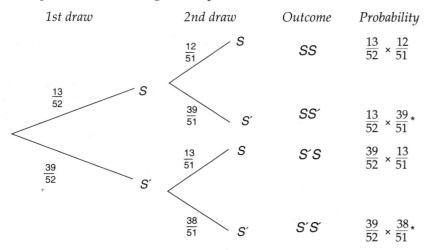

The relevant outcomes are marked * and so:

$$P(B) = \dfrac{13}{52} \times \dfrac{39}{51} + \dfrac{39}{52} \times \dfrac{38}{51} = \dfrac{3}{4}$$

$$\Rightarrow P(A \mid B) = \dfrac{\dfrac{13}{68}}{\dfrac{3}{4}} = \dfrac{13}{51}$$

We met an example earlier where the information about B did not affect the probability of A.

i.e. we had $P(A \mid B) = P(A)$

If this property holds then A is said to be **independent of** B. It is possible to show that if A is independent of B then the converse is also true, i.e. that B is independent of A. This means that we can talk about independent events rather than the independence of one event from another.

So from now on if $P(A \mid B) = P(A)$, we call the events A and B **independent**.

Although the definition of independence involves a statement about a conditional probability, in practice, the following consequence is much simpler to use in problem solving.

$$P(A \mid B) = P(A)$$
$$\Rightarrow \quad \frac{P(A \cap B)}{P(B)} = P(A)$$
$$\Rightarrow \quad P(A \cap B) = P(A) \times P(B)$$

It should be noted that this idea has been used implicitly in several problems in this section. We now have a more precise idea of when multiplying probabilities is legitimate.

The simple multiplication rule stated earlier can only be used when events are **independent**.

In all the situations in which it has so far been used, independence was assumed on the grounds that the events being considered were physically independent. For example, in drawing two cards from a pack with replacement, the result of the first selection can have no influence on the outcome of the second selection. These activities are independent of each other and so the multiplication rule can be used.

To conclude:

> events A and B are independent if: $\quad P(A \cap B) = P(A) \times P(B)$

We can use this both ways. Namely, if we are given that the events are independent, then it is legitimate to multiply their probabilities according to this rule. Conversely, if this rule applies numerically, then we can safely conclude that the events are independent.

Example	Two dice are thrown, one green and the other blue. A is the event 'the green die shows a 3', B is the event 'the sum of the scores is 8'. Are A and B independent?

Solution	$$P(A) = \frac{1}{6} \qquad P(B) = \frac{5}{36}$$ $$A \cap B = \{ (3, 5) \} \Rightarrow P(A \cap B) = \frac{1}{36}$$ $$P(A) \times P(B) = \frac{1}{6} \times \frac{5}{36} = \frac{5}{216} \neq P(A \cap B)$$ Hence the events are dependent.

Example	Given that $P(A) = \frac{1}{3}$, $P(B) = \frac{2}{5}$ and $P(A \cup B) = \frac{3}{5}$, find the values of $P(A \cap B)$, $P(A \mid B)$ and $P(B \mid A)$. Are A and B independent?

Solution	We use $P(A \cup B) = P(A) + (P(B) - P(A \cap B)$ to find $P(A \cap B)$

$$\frac{3}{5} = \frac{1}{3} + \frac{2}{5} - P(A \cap B) \qquad\qquad \Rightarrow P(A \cap B) = \frac{2}{15}$$

$$\text{Now } P(A \mid B) = \frac{P(A \cap B)}{P(B)} = \frac{\frac{2}{15}}{\frac{2}{5}} = \frac{1}{3} \quad \Rightarrow P(A \cap B) = \frac{1}{3}$$

$$P(B \mid A) = \frac{P(B \cap A)}{P(A)} = \frac{\frac{2}{15}}{\frac{1}{3}} = \frac{2}{5} \qquad\qquad \Rightarrow P(A \cap B) = \frac{2}{5}$$

$$P(A) \times P(B) = \frac{1}{3} \times \frac{2}{5} = \frac{2}{15} = P(A \cap B)$$

so A and B are independent.

When A and B are independent, we're familiar with the simple multiplication rule:

$$P(A \cap B) = P(A) \times P(B)$$

When A and B are *not* independent, the conditional probabilities $P(A \mid B)$ and $P(B \mid A)$ are both measures of their dependence on each other. These probabilities can then be used to modify the multiplication rule.

$$P(A \mid B) = \frac{P(A \cap B)}{P(B)}$$

Multiplying both sides by $P(B)$ gives:

$$P(A \mid B) \times P(B) = P(A \cap B)$$

In the same way, $P(B \mid A) = \dfrac{P(A \cap B)}{P(A)} \quad \Rightarrow \quad P(B \mid A) \times P(A) = P(A \cap B)$

$$\therefore \quad P(B \mid A) \times P(B) = P(B \mid A) \times P(A) = P(A \cap B)$$

Example	Given that $P(A) = \frac{2}{5}$ and $P(B \mid A) = \frac{1}{3}$, find $P(A \cap B)$.

Solution	$P(A \cap B) = P(B \mid A) \times P(A) = \frac{1}{3} \times \frac{2}{5} = \frac{2}{15}$.

Practice questions D

1

$$\left(\text{4R 6G} \right) \qquad \left(\text{7R 3G} \right)$$

Two identical looking bags.
The first bag contains 4 red and 6 green discs.
The second bag contains 7 red and 3 green discs.

A bag is chosen at random and a disc is then selected at random.

(a) What is the probability that the disc is red?

(b) Given that the disc chosen is red, what is the probability that it was taken from bag 1?

2 In warehouse A, 40% of the potatoes are rotten. In warehouse B, 30% of the potatoes are rotten. A warehouse is chosen at random and 2 potatoes are then chosen at random.

(a) What is the probability that both potatoes are rotten?

(b) What is the conditional probability that warehouse A was chosen, given that both potatoes are rotten?

3 Farm A: 20% of cows diseased
Farm B: 15% of cows diseased
Farm C: no cows are diseased.

Choose a farm at random. Then select 2 cows at random.

(a) What is the probability that both cows are diseased?

(b) If both cows are diseased, what is the probability you selected from A?

4 48% of a group of people smoke. Those that smoke have a 54% probability of drinking. Those that don't smoke have a 63% probability of drinking.

(a) A person is chosen at random. What is the probability she smokes?

(b) A person is chosen at random. What is the probability he drinks?

(c) A randomly chosen person is found to be a drinker. What is the probability she also smokes?

5 You stand a 60% chance of catching flu. If you get flu, the probability of a sore throat is 85%. If you don't get flu, the probability of a sore throat is 30%. You wake up with a sore throat. What is the probability you've got flu?

6 If $P(A) = \frac{5}{14}$, $P(B) = \frac{5}{7}$ and $P(A \cap B) = \frac{1}{7}$ find:

(a) $P(A')$ (b) $P(A \cup B)$ (c) $P(A \mid B)$.

7 If $P(A) = \frac{1}{2}$, $P(B) = \frac{7}{18}$ and $P(A \cup B) = \frac{2}{3}$ find $P(A \cap B)$.

Are A and B independent? Also find $P(A \mid B)$.

8 If $P(A) = \frac{9}{24}$, $P(B) = \frac{7}{12}$ and $P(A \mid B) = \frac{3}{14}$ find:

(a) $P(A \cap B)$ (b) $P(A \cup B)$ (c) $P(B')$.

9 If $P(A \cup B) = \frac{9}{10}$, $P(A \cap B) = \frac{3}{10}$ and $P(A \mid B) = \frac{3}{5}$ find:

(a) $P(B)$ (b) $P(A)$ (c) $P(A')$ (d) $P(B \mid A)$.

10 If $P(A) = \frac{1}{3}$, $P(B) = \frac{1}{4}$ and $P(A \cup B) = \frac{1}{2}$ find $P(A \cap B)$.

Deduce that A and B are independent.

11 If $P(A) = \frac{1}{3}$, $P(B) = \frac{1}{2}$ and $P(A \cup B) = \frac{3}{4}$ find:

(a) $P(A \cap B)$ (b) $P(A \mid B)$
(c) $P(B \mid A)$ (d) $P(B' \mid A')$
(e) $P(A' \mid B')$ [Hint: $A' \cap B' = (A \cup B)'$].

12 An 8-sided dice with faces marked 1 to 8 is thrown. The following events are defined:
$E_1 = \{1, 2, 3, 4, 5\}$, $E_2 = \{2, 4, 6, 8\}$ and $E_3 = \{1, 3, 5, 7\}$.

Illustrate with a Venn diagram. Find the following:
(a) $P(E_1 \mid E_2)$ (b) $P(E_1 \mid E_3)$ (c) $P(E_1' \mid E_2)$
(d) $P(E_2 \mid E_1)$ (e) $P(E_3 \mid (E_1$ or $E_2))$ (f) $P(E_2 \mid E_3)$.

SUMMARY EXERCISE

1 A and B are two events such that:

$P(A) = 0.6$, $P(B) = 0.2$ and $P(A \cap B) = 0.1$
Find:

(a) $P(A' \cap B)$,

(b) the probability that exactly one of A and B will occur.

2 A die and three coins are simultaneously tossed. Find the probability of the event 'at least one head and a score of more than 4'.

3 An urn contains 3 red, 4 white and 5 blue discs. Three discs are selected at random from the urn. Find the probability that:

(a) all three discs are the same colour, if the selection is with replacement,

(b) all three discs are of different colours, if the selection is without replacement.

4 Container 1 has 3 red balls and 4 blue balls in it. Container 2 has 5 red balls and 2 blue balls in it.

A ball is taken from container 1 and placed in container 2. If a ball is now selected randomly from container 1 what is the probability that it is red?

5 How many three digit numbers can be formed using the digits 1, 2, 3, 4, 5, where each digit can be used only once?

How many of them are odd?

6 In a group of six students, four are female and two are male. Determine how many committees of three members can be formed containing one male and two females.

7 A child has a bag containing 12 sweets of which 3 are yellow, 5 are green and 4 are red. When the child wants to eat one of the sweets, a random selection is made from the bag and the chosen sweet is then eaten before the next random selection is made.

(a) Find the probability that the child does not select a yellow sweet in the first two selections.

(b) Find the probability that there is at least one yellow sweet in the first two selections.

(c) Find the probability that the fourth sweet selected is yellow, given that the first two sweets selected were red ones.

8 A and B are two independent events such that:

$$P(A) = \alpha \quad \text{and} \quad P(A \cup B) = \beta, \; \beta > \alpha$$

Show that:

$$P(B) = \frac{\beta - \alpha}{1 - \alpha}$$

9 In a large group of people it is known that 10% have a hot breakfast, 20% have a hot lunch and 25% have a hot breakfast or a hot lunch. Find the probability that a person chosen at random from this group

(a) has a hot breakfast and a hot lunch,

(b) has a hot lunch, given that the person chosen had a hot breakfast.

10 A and B are two independent events such that $P(A) = 0.2$ and $P(B) = 0.15$.

Evaluate the following probabilities:

(a) $P(A \mid B)$

(b) $P(A \cap B)$

(c) $P(A \cup B)$

11 Team A has a $\frac{2}{3}$ probability of winning whenever it plays.

Given that A plays 4 games, find the probability that A wins more than half of the games.

12 A house is infested with mice and to combat this the householder acquired four cats. Albert, Belinda, Khalid and Poon. The householder observes that only half of the creatures caught are mice. A fifth are voles and the rest are birds.

20% of the catches are made by Albert, 45% by Belinda, 10% by Khalid and 25% by Poon.

(a) The probability of a catch being a mouse, a bird or a vole is independent of whether or not it is made by Albert. What is the probability of a randomly selected catch being a

(i) mouse caught by Albert,

(ii) bird not caught by Albert?

(b) Belinda's catches are equally likely to be a mouse, a bird or a vole. What is the probability of a randomly selected catch being a mouse caught by Belinda?

(c) The probability of a randomly selected catch being a mouse caught by Khalid is 0.05. What is the probability that a catch made by Khalid is a mouse?

(d) Given that the probability that a randomly selected catch is a mouse caught by Poon is 0.2 verify that the probability of a randomly selected catch being a mouse is 0.5.

(e) What is the probability that a catch which is a mouse was made by Belinda?

[AEB 1993]

SUMMARY

● In this section we have used set notation to solve probability questions. In particular, for events A and B:
  – $P(A') = 1 - P(A)$, where A' is the complement of A
  – $P(A \cup B) = P(A) + P(B) - P(A \cap B)$
  – for **conditional probability** we use $P(A \mid B) = \dfrac{P(A \cap B)}{P(B)}$

  where $P(A \mid B)$ stands for the probability of A occuring given that B has already occurred:
  – events are **mutually exclusive** if $P(A \cap B) = 0$.
    In that case $P(A \cup B) = P(A) + P(B)$
  – events are **independent** if:
    $P(A \mid B) = P(A)$ so that $P(A \cap B) = P(A) \times P(B)$.

The solution of other types of probability questions have involved:

● tree diagrams

● permutations $^nP_r = \dfrac{n!}{(n - r)!}$

● combinations $^nC_r = \dbinom{n}{r} = \dfrac{n!}{(n - r)!\, r!}$

ANSWERS

Practice questions A

1

	1	1	1	2	4	4
1	2	2	2	3	5	5
1	2	2	2	3	5	5
1	2	2	2	3	5	5
1	2	2	2	3	5	5
2	3	3	3	4	6	6
2	3	3	3	4	6	6

∴ P(even total) $= \dfrac{18}{36} = \dfrac{1}{2}$

2

	1	2	2	3
1	2	3	3	4
2	3	4	4	5
2	3	4	4	5
3	4	5	5	6

∴ P(total > 3) $= \dfrac{11}{16}$

3 (a) 0.4 (b) 0.4 (c) $P(A \cap C) = 0$
∴ A and C are mutually exclusive.

4 $P(A \cap B) = 0$ and $P(A') = \dfrac{2}{3}$

5 (a) {4,5,6,7,8,9} (b) {4,5,6}
(c) {2,3,4,5,6,7,8} (d) {0,1,9}
(e) {0,1} (f) $\dfrac{6}{10}$ (g) $\dfrac{3}{10}$
(h) $\dfrac{7}{10}$ (i) $\dfrac{3}{10}$ (j) $\dfrac{2}{10}$

Practice questions B

1

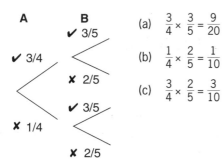

	A	**B**
		✔ 3/5
	✔ 3/4	✘ 2/5
		✔ 3/5
	✘ 1/4	✘ 2/5

(a) $\frac{3}{4} \times \frac{3}{5} = \frac{9}{20}$

(b) $\frac{1}{4} \times \frac{2}{5} = \frac{1}{10}$

(c) $\frac{3}{4} \times \frac{2}{5} = \frac{3}{10}$

2

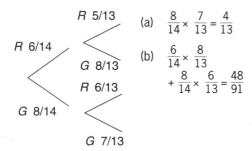

		R 5/13
	R 6/14	G 8/13
		R 6/13
	G 8/14	G 7/13

(a) $\frac{8}{14} \times \frac{7}{13} = \frac{4}{13}$

(b) $\frac{6}{14} \times \frac{8}{13} + \frac{8}{14} \times \frac{6}{13} = \frac{48}{91}$

3 $\frac{8}{20} \times \frac{7}{19} = \frac{14}{95}$

5

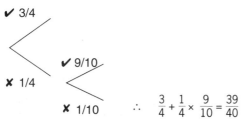

	✔ 3/4	
		✔ 9/10
	✘ 1/4	✘ 1/10

∴ $\frac{3}{4} + \frac{1}{4} \times \frac{9}{10} = \frac{39}{40}$

6 $\frac{11}{12}$

Practice questions C

1 720, 816, 7310, 35, 126

2 360, 90 720

3 45, 495, £64.35

4 336

5 364

6 364, 120

7 16170, 980

8 0.0476, 0.536, 0.0833, 0.0595, 0.238

9 0.0842

10 0.75, 7.6 ∴ 8

Practice questions D

1 (a) $\frac{11}{20}$ (b) $\frac{4}{11}$

2 (a) 0.125   (b) 0.64

3 (a) 0.02083   (b) 0.64

4 (a) 0.48   (b) 0.5868   (c) 0.0442

5 0.81

6 (a) $\frac{9}{14}$ (b) $\frac{13}{14}$ (c) $\frac{1}{5}$

7 (a) $\frac{2}{9}$ (b) No because $\frac{2}{9} \neq \frac{1}{2} \times \frac{7}{8}$ (c) $\frac{4}{7}$

8 (a) $\frac{1}{8}$ (b) $\frac{5}{6}$ (c) $\frac{5}{12}$

9 (a) $\frac{1}{2}$ (b) $\frac{7}{10}$ (c) $\frac{3}{10}$ (d) $\frac{3}{7}$

10 $\frac{1}{12}$

Independent because $\frac{1}{12} = \frac{1}{3} \times \frac{1}{4}$

11 (a) $\frac{1}{12}$ (b) $\frac{1}{6}$ (c) $\frac{1}{4}$ (d) $\frac{3}{8}$ (e) $\frac{1}{2}$

12

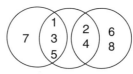

E_3 E_1 E_2

(a) $\frac{1}{2}$ (b) $\frac{3}{4}$ (c) $\frac{1}{2}$ (d) $\frac{2}{5}$ (e) $\frac{3}{7}$ (f) 0

6

Correlation and regression

INTRODUCTION Many situations arise where we might want to investigate whether there is a relationship between two (or more) variables. In this section and the next we will be concerned with the possible links between two variables (bivariate data).

By the end of this section you should be able to:

● plot and interpret a scatter diagram

● determine how well the data are correlated by calculating the product moment correlation coefficient

● fit lines of best fit (known as regression lines)

● find the equations of these regression lines and be able to use and interpret them.

Scatter diagrams

OCR **S1** 5.11.4 (c)

Consider the data in Table 6.1

Table 6.1

Item	1	2	3	4	5	6	7	8	9	10
Height (cm)	158.8	172.1	154.8	172.8	172.9	168.8	171.3	159.9	162.8	164.8
Weight (kg)	48.9	100.7	37.2	90.4	88.1	80.7	79.6	68.1	53.2	65.0

We would like to know whether a relationship exists between the two variables, height and weight. Common sense suggests that there might be, but we would like more evidence than just our intuitive feelings about the situation. A good rough and ready way is to plot the data in graphical form and we do this by considering each item to be an ordered pair and then plotting the graph in the normal way. This is shown in Fig. 6.1 and is called a scatter diagram.

Since each item has two measurements associated with it, we call this a **bivariate** situation.

Figure 6.1

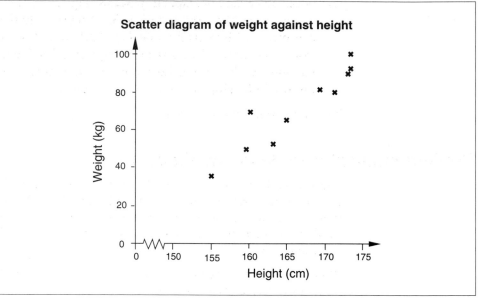

There are a number of points that we could make about the scatter diagram:

- It is important to choose scales and axes so that the data is clearly shown. It would not be sensible in this example to start the axis at zero.

- The two variables appear to increase together: larger values for height seem to be associated with larger values for weight.

- There is a suggestion of some sort of linear relationship between the two variables.

In Fig. 6.2 some further common types of scatter diagrams are shown.

Figure 6.2

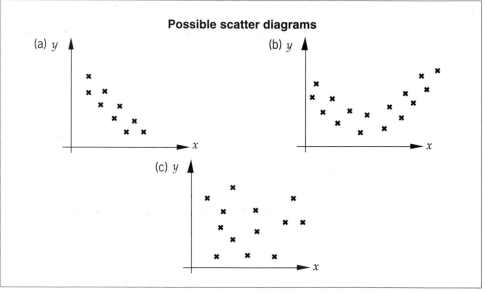

A possible interpretation of each of these might be:

(a)   perhaps a linear relationship between x and y, with y decreasing as x increases (i.e. negative gradient)

(b) not a linear relationship, but possibly a quadratic one between x and y (i.e. y is possibly some quadratic function of x). In this module we are only concerned with linear relationships (although it is possible to transform from non-linear to linear relationships by using logarithms or changes of variable). In the case of the quadratic relationship $y = kx^2$ there would be a linear relationship between $Y = y$ and $X = x^2$.

(c) no relationship observable.

Product moment correlation coefficient

OCR **S1** 5.11.4 (a),(c)

The scatter diagram gives a good general impression of the data and will usually be the first step in interpreting it. It is also valuable to have a numerical measure of the degree of linearity of the relationship – this is provided by the **product moment correlation coefficient** (usually referred to more simply as the correlation coefficient).

What we need is an expression which will give us 1 (or 100%) for a situation such as the one shown in Fig. 6.3:

Figure 6.3

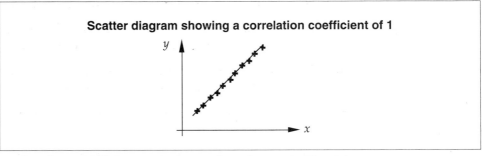

Scatter diagram showing a correlation coefficient of 1

and –1 (or –100%) for a situation such as the one in Fig. 6.4:

Figure 6.4

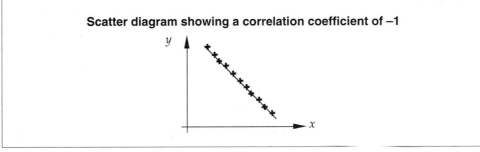

Scatter diagram showing a correlation coefficient of –1

and (say) 0.9 for the one in Fig. 6.5:

Figure 6.5

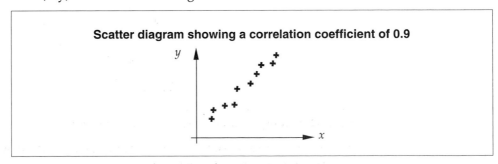

Scatter diagram showing a correlation coefficient of 0.9

and (say) –0.4 for the one in Fig. 6.6:

Figure 6.6

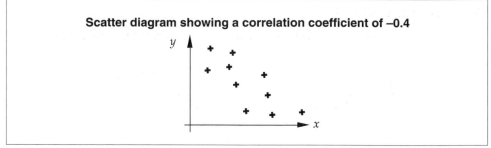

Scatter diagram showing a correlation coefficient of –0.4

The correlation coefficient (written r) which does this is given by:

$$r = \frac{\sum x_i y_i - n\bar{x}\,\bar{y}}{\sqrt{(\sum x_i^2 - n\bar{x}^2)(\sum y_i^2 - n\bar{y}^2)}}$$

and, in all situations $-1 \le r \le 1$.

When $r = 1$ we say that all the data have perfect **positive** linear correlation.

When $r = -1$ we say that all data have perfect **negative** linear correlation.

Let's see how this works out with the data given by Table 6.1.

We have:

Item	(x) Height (cm)	(y) Weight (kg)	x^2	y^2	xy
1	158.8	48.9	25217.44	2391.21	7765.32
2	172.1	100.7	29618.41	10140.49	17330.47
3	154.8	37.2	23963.04	1383.84	5758.56
4	172.8	90.4	29859.84	8172.16	15621.12
5	172.9	88.1	29894.41	7761.61	15232.49
6	168.8	80.7	28493.44	6512.49	13622.16
7	171.3	79.6	29343.69	6336.16	13635.48
8	159.9	68.1	25568.01	4637.61	10889.19
9	162.8	53.2	26503.84	2830.24	8660.96
10	164.8	65.0	27159.04	4225.0	10712.00
	1659.0	711.9	275621.16	54390.81	119227.75

$$\therefore \bar{x} = \frac{1659.0}{10} = 165.9, \qquad \bar{y} = \frac{711.9}{10} = 71.19$$

and

$$\sum x^2 = 275621.16, \qquad \sum y^2 = 54390.81, \qquad \sum xy = 119227.75$$

$$\therefore r = \frac{119227.75 - 10 \times 165.9 \times 71.19}{\sqrt{(275621.16 - 10 \times 165.9^2)(54390.81 - 10 \times 71.19^2)}}$$

$$= 0.9303 \ (4 \ \text{d.p})$$

i.e. the data shows a positive 93.03% linear correlation.

This supports what was originally showed by the scatter diagram.

Practice questions A

1 Work out the product moment correlation coefficient for the following sets of data. Comment each time.

 C 3.2

(a)

x	2	3	4	5	7
y	3	5	8	8	9

(b)

x	3	5	7	11
y	4	8	12	20

(c)

x	3	4	5	7	9
y	8	7	6	3	2

(d)

x	3	4	5	9
y	7	2	5	3

2 Draw a scatter diagram to represent the following data:

 C 3.2

x	1	2	3	4	5
y	5	3	3	2	1

Describe any correlation shown.

Calculate the product moment linear correlation coefficient.

Comment on its value.

This is all very well, but what a long and tedious job it is! Luckily *the calculator comes to our rescue* and does all the hard work for us.

The usual procedure is as follows:

- Put LR mode on
- Shift AC
- Feed in as x [(--- y M+

When this is done:

- K out 3 (this checks that n is correct)
- Shift 9 (this gives the correlation coefficient)

Other models proceed as follows:

- Puft LR mode on Mode 3 1
- Shift AC =
- Feed in a x • y M+

When this is done:

- Rcl C (this checks that n is correct)

● $\boxed{\text{Shift}}$ $\boxed{r}$ $\boxed{=}$ (this gives the correlation coefficient)

It is *very important* that your calculator has an LR mode and that you know how to use it.

Practice questions B

1 Use your calculator to find the product moment correlation coefficient for the following sets of data.

(a)

x	0	1	2	3	4	5
y	3	8	9	11	14	17

(b)

x	0	2	4	6	8	10
y	5	6	8	9.2	11.8	12.4

(c)

x	1.3	2.5	3.8	4.2	8.9
y	0.1	0.3	0.7	1.1	1.8

(d)

x	2.1	3.2	4.5	5.8	6.2
y	8.3	7.6	7.2	6.9	6.3

Sometimes the data are given in summarised form.

Example

$n = 10$, $\sum x = 61$, $\sum x^2 = 521$, $\sum y = 106$, $\sum y^2 = 1424$, $\sum xy = 851$.

Use your calculator to find the correlation coefficient.

Solution

The process now is to $\boxed{\text{K in}}$ the data first of all.

It is usual to proceed as follows:

● Put LR mode on

● Shift AC

Feed in as:

● 10 $\boxed{\text{K in}}$ $\boxed{3}$

● 61 $\boxed{\text{K in}}$ $\boxed{2}$

● 521 $\boxed{\text{K in}}$ $\boxed{1}$

● 106 $\boxed{\text{K in}}$ $\boxed{5}$

● 1424 $\boxed{\text{K in}}$ $\boxed{4}$

● 851 $\boxed{\text{K in}}$ $\boxed{6}$

Then $\boxed{\text{Shift}}$ $\boxed{9}$ will give the correlation coefficient $r = 0.966 \ldots$

(The other models you *store* $\sum x^2$, $\sum x$, n, $\sum y^2$, $\sum y$, $\sum xy$ respectively in A, B, C, D, E and F respectively. Once again, make sure you can really use your calculator!)

Practice questions C

1 Find the product moment correlation coefficient for the following:

(a) $n = 9$, $\sum x = 54$, $\sum x^2 = 384$, $\sum y = 240$, $\sum y^2 = 6582$, $\sum xy = 1342$

(b) $n = 12$, $\sum x = 73$, $\sum x^2 = 653$, $\sum y = 97$, $\sum y^2 = 1219$, $\sum xy = 890$

(c) $n = 5$, $\sum x = 21$, $\sum x^2 = 103$, $\sum y = 20$, $\sum y^2 = 106$, $\sum xy = 103$

2 The height h and weight w of 10 people are measured:

$\sum h = 1710$, $\sum w = 760$, $\sum h^2 = 293162$, $\sum hw = 130628$, $\sum w^2 = 59390$

Find the product moment correlation coefficient.

Once we've established that correlation exists, the next thing we have to do is to find the equation of the best fit line (or lines). These lines are called **regression lines** but before we can calculate them, we need to consider the general situation.

Linear regression

OCR **S1** 5.11.4 (e),(g),(h)

Examples of situations where a linear model may be appropriate are:

(a) Does increasing the amount of fertiliser applied to a certain crop increase the yield?

(b) Does attendance at a training course and subsequent examination improve the performance of salespeople working for a company?

(c) Is there a linear relationship between the marks obtained in Paper I and those obtained in Paper II of an examination?

In the first two of these examples there is a degree of asymmetry in the relationship.

In example (a), for instance, we could conduct a controlled experiment in which we decide on the amount of fertiliser to apply (X) and then observe the change in yield (Y).

In example (b) we can test the relationship between the outcome of the exam (X) against the change in performance of the salespeople (Y).

In each of these cases we refer to the variable X as the **independent** or **explanatory** variable. It is something we have control over. The values of the variable Y are what we observe in response to changes in the value of X and the Y-values are consequently referred to as **dependent** or **response** variables.

In example (c) it is not so clear which variable should be called independent and which should be called dependent.

We will now look at the problem of finding a line of best fit by considering an example in some detail.

Number of items (x) (1000 s)	21	39	48	24	72	75	15	35	62	81	12	56
Production cost (y) (£ 1000)	40	58	67	45	89	96	37	53	83	102	35	75

(a)   Plot these data on a scatter diagram. Explain why this diagram would support the fitting of a regression equation of y on x.

(b)   Find an equation for the regression line of y on x in the form $y = a + bx$.

(Use $\sum x^2 = 30\,786$; $\sum xy = 41\,444$)

The selling price of each item produced is £2.20.

(c)   Find the level of output at which total income and total costs are equal. Interpret this value.

| Solution | (a)   Fig. 6.7 shows the data plotted on a scatter diagram. |

Figure 6.7

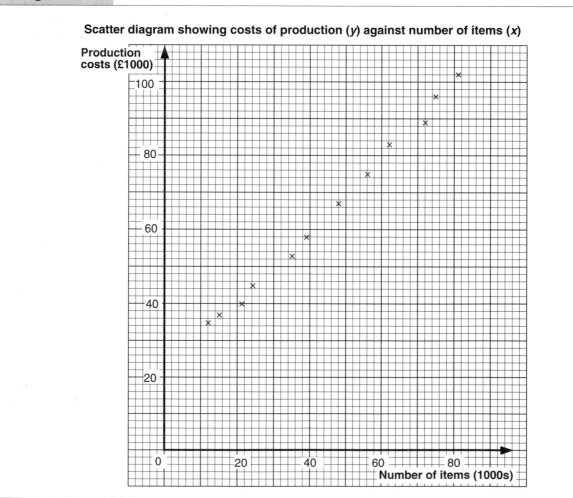

Scatter diagram showing costs of production (*y*) against number of items (*x*)

The points lie fairly closely on a straight line and so we conclude that it is worthwhile finding a regression line.

A quick calculation shows that the correlation coefficient between x and y is 0.99 which provides strong evidence for linearity when combined with the visual evidence.

(b) In this example there is a clear dependency of the variable y on the variable x. The value of x is given (it is the amount of production per month) and it is this that determines the costs for that month. So x is our explanatory variable and y is our response variable.

This part of the question asks us to find the equation of the regression line of y on x and this means finding a formula

$$y = \alpha + \beta x$$

i.e. to give y as a linear function of x.

[Later we will see that it is also possible to find a regression line of the form $x = c + dy$ which is called the regression line of x on y. In general these lines will be different.]

The problem is now reduced to using the values of the sample data to find the appropriate values of α and β. 'A' level examining boards provide the formulae necessary for calculating α and β and so it is not necessary to remember them.

The formula for the regression line of y on x is

$$y = \alpha + \beta x$$

where

$$\beta = \frac{\sum x_i y_i - n \bar{x}\, \bar{y}}{\sum x_i^2 - n \bar{x}^2}$$

and $\alpha = \bar{y} - \beta \bar{x}$

A consequence of the derivation of these formulae is that **the regression lines for y on x and for x on y both pass through the point $(\bar{x}, \bar{y})$**, and that in the case of perfect correlation ($r = +1$ or $r = -1$), the lines would actually coincide.

The point $(\bar{x}, \bar{y})$ can now be indicated on the scatter diagram (as in Fig. 6.8) and the regression line drawn through it.

It is worth pointing out here that if we were calculating the regression of x on y, the same formulae would be used, except that the x's and y's would be interchanged, so that for:

$$x = c + dy$$

we would have:

$$d = \frac{\sum x_i y_i - n\bar{x}\,\bar{y}}{\sum y_i^2 - n\,\bar{y}^2}$$

$$c = \bar{x} - d\bar{y}$$

i.e. there is complete symmetry in the formulae.

Which line we find, i.e. either y on x or x on y (or possibly both), depends on what it is we want to know. If we wish to predict y-values from the regression line, then it is y on x that is appropriate and for predicting x-values, then we use the regression line of x on y. If both types of prediction are required, then it is necessary to find both of the regression lines.

As in many problems of this type, to avoid tedious calculations in examination conditions, we are provided with some summary statistics, namely

$$\sum x^2 = 30\ 786 \text{ and } \sum xy = 41\ 444$$

We also find from the table that $\bar{x} = 45$, $\bar{y} = 65$.

Using these directly in the formula for β, we get

$$\beta = \frac{(41\ 444) - (12)\ (45)\ (65)}{(30\ 786) - (12)\ (45)^2}$$

$$= \frac{6344}{6486} = 0.98 \text{ (2 d.p.)}$$

and $\alpha = 65 - (0.98)\ (45) = 20.99$ (2 d.p.)

The line of best fit is therefore

$$y = 20.99 + 0.98\ x$$

and this now plotted on the scatter diagram (see Fig. 6.8).

The simplest way to plot the line is to observe that the intercept on the y-axis is 20.99 and that the line must pass through $(\bar{x}, \bar{y})$.

Alternatively, if the intercept value is not shown on the vertical axis, e.g. if it is negative or if we have used axes with 'breaks' in them, then we could just find two points on the line or use one point and $(\bar{x}, \bar{y})$.

(c)   Ignoring the fact that everything is multiplied by 1000 in the units, if x items are produced

profit = £2.20 × x

But the total cost for producing x items is

$$y = 20.99 + 0.98x$$

If these are to be equal then we require

$$2.20x = 20.99 + 0.98x$$

$$\Rightarrow \quad 1.22x = 20.99$$

$$\Rightarrow \quad x = 17.2049$$

Therefore we would produce 17 205 items

This point represents the break-even point when costs of production are equal to total income from the production.

Figure 6.8

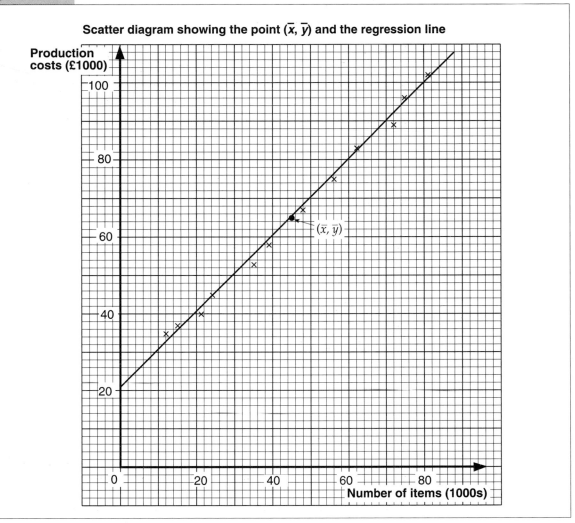

Scatter diagram showing the point $(\bar{x}, \bar{y})$ and the regression line

---

Practice questions D

1 For each of the following find the regression line of y on x.

(a)

x	2	4	5	6	8
y	1	4	9	10	12

$(\sum x = 25, \sum x^2 = 145, \sum xy = 219, \sum y = 36)$

(b) $n = 6$, $\sum x = 22$, $\sum x^2 = 100$, $\sum xy = 144$, $\sum y = 30$.

2 The number x of oak trees and the ground moisture content y are found in each of 10 equal areas. The following is a summary of the survey:

$\sum x = 500$, $\sum y = 300$, $\sum x^2 = 27818$, $\sum xy = 16837$, $\sum y^2 = 10462$

Find the regression line of y and x.

Estimate the moisture content in an area which contains 60 oak trees.

Find also the product moment correlation coefficient.

3 For

x	31	33	34	35	38
y	23	24	27	28	33

($\sum x = 171$, $\sum y = 135$, $\sum x^2 = 5875$, $\sum xy = 4657$)

Find the regression line of y on x

Hence estimate y when $x = 35.2$

You will have probably realised by now that *the calculator can do all this work for you*. All you need to know, once you've fed in the data, is how to key out values for α and β.

It is usual to:

- Shift 7 to find α

- Shift 8 to find β.

and, of course, Shift 9 to find r.

In other models:

- Shift A = to find α

- Shift B = to find β

and, of course, Shift r = to find r.

There is no need to show any working but, of course, you must get it right! You need to be able to use your calculator efficiently.

Practice questions E

1 Use your calculator, for each of the following sets of data, to find:
  (i)   product moment correlating coefficient
  (ii) regression line of y and x
  (iii) the point $(\bar{x}, \bar{y})$.

(a)
x	3	5	5	6	8	9	9	10	10	12
y	1	4	2	5	7	7	9	9	8	12

(b)
x	13	14	15	16	17	18
y	0.2	0.2	0.4	0.6	0.6	1

We will now work through another two complete 'A' level questions more concisely to illustrate how such solutions should be written up.

Example

Three trainee technicians carry out laboratory trials to examine the effect of temperature on the yield of an industrial process. The table shows the results obtained by each technician.

Technician	A	B	C	A	B	C	A	B	C	A	B	C
x, temperature, °C	10	15	20	25	30	35	40	45	50	55	60	65
y, yield, kg	80	106	75	90	117	118	97	127	80	109	140	115

$\sum x = 450$ $\sum y = 1254$ $\sum x^2 = 20\,450$ $\sum xy = 49\,245$

(a) (i)   Draw a scatter diagram of the data. Label each point A, B or C according to which technician carried out the trial.

    (ii)   Calculate the equation of the regression line of yield on temperature and draw the line on your scatter diagram

    (iii)   Use your equation to estimate the yield for a temperature of 52°C

It is known that over this range of temperatures the relationship between yield and temperature is approximately linear.

(b)   Comment on the performance of the three trainee technicians and on the reliability of the estimate made in (a) (iii).

An experienced and reliable technician carries out the trial at a temperature of 40°C and obtains a yield of 125 kg.

(c)   Plot this point on your scatter diagram. Without further calculation modify the estimate made in (a) (iii). Comment on the reliability of your new estimate.

Solution

(a) (i)   The scatter diagram is shown in Fig. 6.9.

    (ii)   Finding regression of y on x:

$$\bar{x} = 37.5, \quad \bar{y} = 104.5$$
$$\beta = \frac{(49\,245) - (12)\,(37.5)\,(104.5)}{(20\,450) - (12)\,(37.5^2)} = 0.62 \text{ (2 d.p.)}$$
$$\alpha = 104.5 - \beta\,37.5 = 81.25 \text{ (2 d.p.)}$$

giving

$$y = 81.25 + 0.62x.$$

(Alternatively, *write this down* directly from your calculator.)

This line is now plotted on the scatter diagram (see Fig. 6.9).

    (iii)   If $x = 52$°C then from the formula for the equation of the line

$$y = 113.5 \text{ kg}$$

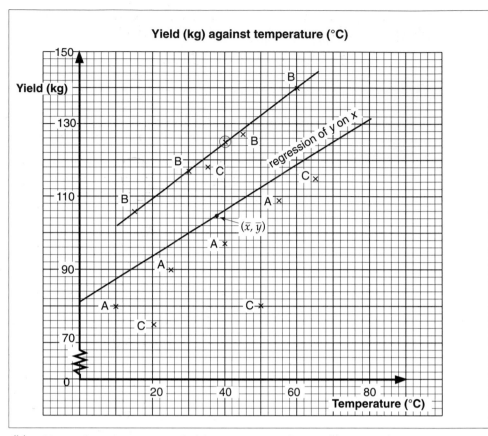

Figure 6.9

Yield (kg) against temperature (°C)

(b)   *C*'s performance is unreliable. His points do not show any linearity.

A and *B* are both linear although they give different lines and there appears to be a constant difference between their results. We have no additional information to say which is the more accurate of *A* and *B*.

The uncertainty described makes the prediction in (a)(iii) unreliable

(c)   Point (40, 125) is plotted as $\otimes$ and it is now apparent that *B*'s results are more consistent with the experienced technician. A line of best fit (by eye) through the points relating to *B* and the experienced technician's point gives the upper line shown.

Using this line our modified estimate is $y = 134$ kg when $x = 52°C$.

Example

Marks out of 100 on Paper I and Paper II for a science exam are given in the following table. Draw a scatter diagram of the data and find the equations of the regression lines of y on x and x on y.

Use these equations to determine estimated marks for Candidate *J* on Paper I and for candidate *D* on Paper II who were absent for these respective papers.

	A	B	C	D	E	F	G	H	I	J	K	L	M	N	O	
Paper I	63	72	38	52	87	63	41	52	68	Abs	45	72	58	58	77	x
Paper II	58	59	28	Abs	85	67	48	45	64	74	53	78	51	65	61	y

Solution	The scatter diagram is shown in Fig. 6.10.

The labelling of x and y is arbitrary

$$\bar{x} = \frac{794}{13} = 61.1, \; \bar{y} = \frac{762}{13} = 58.6, \; \sum x^2 = 51\,010, \; \sum y^2 = 47\,208, \; \sum xy = 48\,668$$

(These values have been calculated *omitting the marks for D and J*.)

Figure 6.10	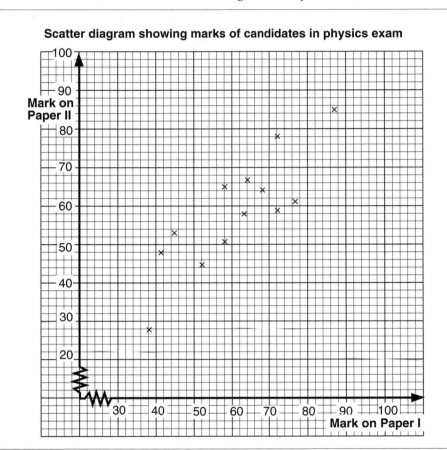

Scatter diagram showing marks of candidates in physics exam

For regression of y on x

$$\beta = \frac{48\,665 - (13)\,(61.1)\,(58.6)}{51\,010 - (13)\,(61.1)^2} = \frac{2122.02}{2478.27} = 0.86$$

$$\alpha = 58.6 - 0.86 \times 61.1 = 6.05$$

giving $y = 6.05 + 0.86x$

(or simply write this down directly from your calculator)

This line passes through $(\bar{x}, \bar{y})$ i.e. (61.1, 58.6) and (40, 40.45).

For regression of x on y

$$\beta = \frac{48\,668 - (13)\,(61.1)\,(58.6)}{47\,208 - (13)\,(58.6)^2} = \frac{2122.02}{2566.52} = 0.83$$

$$\alpha = 61.1 - (0.79)\,(58.6) = 12.46$$

giving $x = 12.46 + 0.83y$

(To obtain this directly from your calculator feed in the other way round.)

This line passes through $(\bar{x}, \bar{y})$ i.e. (61.1, 58.6) and (80, 78.86).

Fig. 6.11 now shows the point $(\bar{x}, \bar{y})$ and the regression lines added to the scatter diagram of Fig. 6.10.

Figure 6.11

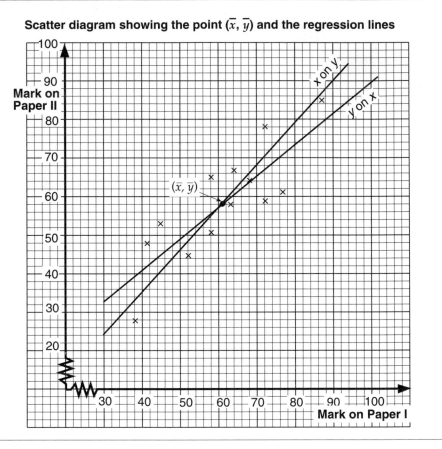

Scatter diagram showing the point $(\bar{x}, \bar{y})$ and the regression lines

We can now use either the equations or the lines on the graph to find estimates of the missing marks.

For candidate D we want his y-value when the x-value is 52. We use the regression of y on x for this giving

Mark on Paper II for D is 51.

For candidate J, we want his x-value when the y-value is 74. Use the regression of x on y for this to give

Mark on Paper I for J is 75.

Practice questions F

1   Consider the data shown below:

x	2	3	5	6
y	1	7	6	12

Find:

- the product moment correlation coefficient
- the regression line of y on x
- the values of $\bar{x}$ and $\bar{y}$
- an estimate of y when $x = 8$
- the regression line of x on y
- the coordinates of intersection of the regression lines of y on x and x on y
- an estimate of x when $y = 9$.

2   Consider the data shown below:

x	5	3	9	1	6
y	9	10	3	12	7

Find:
- the product moment correlation coefficient
- the regression line of y on x
- the regression line of x on y
- an estimate of y when $x = 10$
- an estimate of x when $y = 12$.

Outliers

OCR **S1** 5.11.4 (h)

It may be that bivariate data which is collected follows an apparently linear pattern, apart from one or two exceptional cases which distort the general picture. These will sometimes occur at low or high values of x or y. It may be that these **outliers**, as they are called, are due to limitations on the linear model. In these circumstances it would be wise to reject the linear model in that region (it may still be valid elsewhere) and not use these points in calculations of α and β.

An example is shown in Fig. 6.12.

Figure 6.12

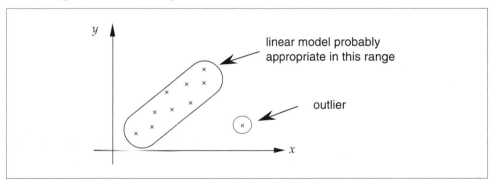

Bizarre readings within a region of suspected linearity should be treated with caution. For example:

Figure 6.13

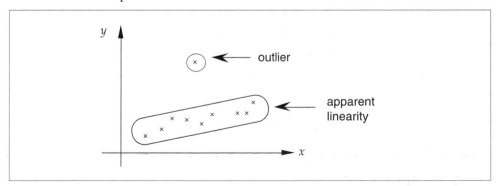

It may be that an error was made in recording that particular data item and if possible, this should be checked. It may be a genuine reading, in which case, if it is used in calculating α and β, it will distort the line of regression away from the more obviously linearly related items and thereby reduce the value of the line. Probably the wisest course of action would be to collect some more data in the region of the outlier value and confirm that it is just an isolated value. If it is not, then the linearity model will probably not be justified at all.

A caution

In some examples of linear regression there is a natural limit to the estimation process using lines of regression, e.g. in the earlier example on exam marks, there would be no call to estimate the mark on Paper II for somebody scoring 105 on Paper I since the possibility simply doesn't exist. We are confined to marks between 0 and 100 in this example.

However, in the example before, concerning the effect of temperature on yield of an industrial process, we ought to be hesitant about extrapolating much beyond the range given for the temperature in the initial data.

We would probably be safe to use B's values for predictions within the interval 0°–70° and possibly a little beyond, but would we want to use this line to make a prediction of yield at a temperature of, say, 250°C? Probably not – since it may be that other factors come into play at higher temperatures. We may experience 'diminishing returns', or our product may have become gaseous at that temperature. We just don't have evidence about temperatures that high and so we shouldn't make such unsafe predictions.

Similarly in the first example in this section, which was concerned with production costs, we would need to exercise caution about how the relationship continues outside the existing ranges for x and y.
The relationship between costs of production and output may cease to be linear outside the range given due to 'economies of scale'.

Practice questions G

1 Name

Name	A	B	C	D	E	F
x	2	6	3	10	0	6
y	8	6	5	2	2	3

  ● Illustrate the data with a scatter diagram.
  ● Who is the outlier?
  ● Work out the product moment correlation coefficient but *without* including the outlier.

  ● Again without the outlier find
  (a) the regression line of y on x
  (b) the regression line of x on y.
  ● Draw in these lines on your scatter diagram.
  ● Use your lines to estimate:
  (c) y when $x = 11$
  (d) x when $y = 16$. **C3.2**
  ● Comment on the validity of the estimates given in (c) and (d).

Interpreting the equation of regression of *y* on *x*

When we were considering our production costs example, we finished up with a regression line of $y = 20.99 + 0.98x$, i.e.

$$\text{Production costs} = 20.99 + 0.98 \times \text{Number of items.}$$

But what interpretation can we give to 20.99 and 0.98?

● 20.99 is the cost of producing no items, i.e. the standing cost

● 0.98 gives the rate at which the production costs are increasing, i.e. for every additional item, the production costs increase by 0.98.

It is important that you are able to make such interpretations.

Practice questions H

1 The regression line for our second example was: **C** 3.2

Yield = 81.25 + 0.62 × Temperature

Interpret the coefficients in this regression line.

2 The regression line for our third example was: **C** 3.2

Paper II marks = 6.05 + 0.86 × Paper I marks.

Interpret the coefficients in this regression line.

Correlation 'difficulties'

Sometimes a non-linear model might be appropriate but, with skill, we might be able to change this into a linear model. Let's look at an example.

| Example |

We are given:

x	1	2	3	4
y	2	14	33	60

It is thought that the model relating y to x could be of the form:

$$\frac{y}{x} = a + bx$$

Find the regression line of $\frac{y}{x}$ on x and hence estimate y where $x = 5$.

| Solution |

We begin by setting up an appropriate table:

x	1	2	3	4
$\frac{y}{x}$	2	7	11	15

Feeding this data into the calculator gives:

$\beta = 4.3$ and $\alpha = -2$

so that the best fit curve will be

$$\frac{y}{x} = -2 + 4.3x$$

And $x = 5 \Rightarrow \frac{y}{5} = -2 + 4.3 \times 5 \Rightarrow y = 97.5$.

Practice questions I

1 You are given:

x	1	2	3	4	5
y	1	3	9	20	40

It is thought that the model relating y to x could be of the form $\frac{y}{x} = a + bx^2$.

Find the regression line of $\frac{y}{x}$ on x^2 and hence estimate y when $x = 6$.

2 You are given:

x	1.7	2.8	3.2	3.0	3.7
y	1.0	1.5	2.0	2.5	4.0

Find the regression line of y on $\frac{1}{x}$ and hence estimate y when $x = 5$.

SUMMARY EXERCISE

1 Cucumbers are stored in brine before being processed into pickles. Data were collected on x, the percentage of sodium chloride in the salt used to make brine, and y, a measure of the firmness of the pickles produced. The data are shown below:

x	6.0	6.5	7.0	7.5	8.0	8.5	9.0	9.5
y	15.3	15.8	16.1	16.7	17.4	17.8	18.2	18.3

you may assume that

$$\Sigma x = 62 \qquad \Sigma x^2 = 491$$
$$\Sigma xy = 1060.6 \qquad \Sigma y = 135.6$$

(a) Plot a scatter diagram of the data. Choose your scale so that values of x up to 12 will fit on the diagram.

(b) Calculate the equation of the regression line of y on x and plot the line on your diagram.

(c) Use the equation of the line to predict the value of y when x is

(i) 6.7 (ii) 10.7

Comment on these predictions.

(d) Further trials were carried out with an increased percentage of sodium chloride in the salt and the following additional observations were obtained.

x	10.0	11.0	12.0
y	18.4	18.2	18.3

Add these points to your original scatter diagram. Modify where appropriate the predictions made in (c) and comment on your new predictions.

(*No further calculations are required for this part*)

[AEB]

2 An electric fire was switched on in a cold room and the temperature of the room was noted at five minute intervals.

Time, minutes, from switching on fire, x	0	5	10	15	20	25	30	35	40
Temperature, °C, y	0.4	1.5	3.4	5.5	7.7	9.7	11.7	13.5	15.4

You may assume that

$$\Sigma x = 180 \qquad \Sigma y = 68.8$$
$$\Sigma xy = 1960 \qquad \Sigma x^2 = 5100$$

(a) Plot the data on a scatter diagram.

(b) Calculate the regression line $y = a + bx$ and draw it on your scatter diagram.

(c) Predict the temperature 60 minutes from switching on the fire. Why should this prediction be treated with caution?

(d) Starting from the equation of the regression line $y = a + bx$, derive the equation of the regression line of

(i) y on t where y is temperature in °C (as above) and t is time in hours

(ii) z on x where z is temperature in °K and x is time in minutes (as above)

(A temperature in °C in converted to °K by adding 273, e.g. 10°C = 280°K.)

(e) Explain why, in (b) the line $y = a + bx$ was calculated rather than $x = a' + b'y$.

If, instead of the temperature being measured at 5 minute intervals, the time for the room to reach predetermined temperatures (e.g. 1, 4, 7, 10, 13°C) had been observed, what would the appropriate calculation have been? Explain your answer.

[AEB]

3 (a) State, with a reason in **each** case, but *without doing any detailed calculations*, whether or not

$$y = -1.4 + 1.6x$$

could be the least squares regression line for either of the following data sets.

(i)

x	3.7	4.4	5.6	6.2	7.3
y	4.5	5.6	7.6	8.5	10.3

(ii)

x	0.3	1.1	1.4	1.8	2.3
y	1.2	0.1	−0.4	−0.9	−1.6

(b) In an investigation of the effect of ammonia on the survival of rainbow trout reared in an intensive static-water environment, the survival rate, $y\%$, was measured at eleven fixed levels of ammonia exposure, x mg/l. The following summarised quantities were then calculated.

$$\Sigma x = 220 \qquad \Sigma y = 860 \qquad \Sigma xy = 16\,546$$
$$\Sigma x^2 = 4840 \qquad \Sigma y^2 = 68\,260$$

State, with a reason, an analysis you would perform on these data *prior* to a regression analysis.

Assuming that this preliminary analysis gives satisfactory results, explain why a regression line of y on x, but **not** of x on y, is appropriate.

Calculate the equation of the least squares regression line of y on x.

Hence estimate the survival rate if the level of ammonia exposure is 25 mg/l. [AEB]

4 The following table shows the amount of water, in cm, applied to seven similar plots on an experimental farm. It also shows the yield of hay in tonnes per acre.

Amount of water (x)	30	45	60	75	90	105	120
Yield of hay (y)	4.85	5.20	5.76	6.60	7.35	7.95	7.77

(Use $\sum x^2 = 45675$; $\sum y^2 = 304.8980$; $\sum xy = 3648.75$)

(a) Find the equation of the regression line of y on x in the form $y = a + bx$.

(b) Interpret the coefficients of your regression line.

(c) What would you expect the yield to be for $x = 28$ and for $x = 150$? Comment on the reliability of each of your expected yields. **C 3.2**

5 A farm food supplier monitors the number of hens kept, x, against the weekly consumption of food, y kg, for a sample of 10 small holdings.

The results are summarised below.

$\sum x = 360$, $\sum x^2 = 17362$

$\sum y = 286$, $\sum y^2 = 10928.94$,

$\sum xy = 13773.6$

(a) Obtain the regression equation for y on x in the form $y = a + bx$.

(b) Give a practical interpretation to the slope b. **C 3.2**

(c) If food costs £7.50 for a 25 kg bag, estimate the weekly cost of feeding 48 hens

6 An experiment was conducted into the effect of density of planting x (in plants/m^2) on the yield per plant y (in grams) of onions. Twelve observations were made as follows:

x	106.53	48.66	35.76	80.73	46.82	45.34
y	61.84	131.27	147.77	76.63	116.36	128.70

x	69.30	98.05	59.35	53.45	67.09	63.04
y	88.55	56.61	94.94	115.12	85.73	93.64

(a) Plot the points on a suitably labelled graph.

(b) Obtain the regression line of y on x in the form $y = a + bx$

(c) Give a practical interpretation to the values of a and b from such a regression line, and comment on how the value of a can be interpreted in the present case. **C 3.2**

(d) Calculate the value of x for which the value of y in the regression line would reach 0, the origin. Comment on the practical value of your result.

(e) Superimpose your regression line on the plot done in part (a). Hence discuss briefly whether a straight line appears to be an adequate model for the relationship between yield and density, and suggest an alternative, more refined model.
[$\sum x = 774.12$, $\sum x^2 = 55031.10$, $\sum y = 1197.16$, $\sum xy = 70858.91$]

7 The sales manager of a large retailer of electrical appliances is monitoring the effects of a radio advertising campaign. Over the last seven weeks differing amounts of radio time x, in minutes, have been purchased and the corresponding numbers of sales y, in hundreds of appliances, have been recorded for the same weeks.

x (minutes)	15	8	22	11	25	18	20
y (hundreds)	16	11	20	15	26	32	20

Plot these data on a scatter diagram.

Calculate, to 2 decimal places, the value of the product moment correlation coefficient.

($\sum x^2 = 2243$; $\sum xy = 2559$; $\sum y^2 = 3102$)

Using your scatter diagram and the analysis carried out so far, explain why it might be inadvisable to fit a straight line regression model to these data.

In the light of your previous answer, select six points from the given data to which you could fit a straight line regression model. Explain your choice of points. **C 3.2**

Find an equation of the regression line of y on x, giving the value of the coefficients to 2 decimal places. Give an interpretation of the slope of your line.

SUMMARY

In this section we have:

- plotted **scatter diagrams** and interpreted them

- evaluated the **product moment correlation coefficient** r by either using the formula

$$r = \frac{\sum x_i y_i - n\bar{x}\,\bar{y}}{\sqrt{(\sum x_i^2 - n\bar{x}^2)(\sum y_i^2 - n\bar{y}^2)}}$$

  or by using our calculator directly

- seen that $-1 \leq r \leq 1$, with $r = 1$ implying **perfect positive linear correlation** and $r = -1$ **perfect negative linear correlation**

- worked out the **regression line of y on x** by either using the formula

$$y = \alpha + \beta x \qquad \text{where } \alpha = \bar{y} - \beta\bar{x} \text{ and } \beta = \frac{\sum xy - n\bar{x}\,\bar{y}}{\sum x^2 - n\bar{x}^2}$$

  or by using our calculator directly.

- **used the regression line of y on x** to predict y given x (but only over the given range of x, or thereabouts)

- worked out the **regression line of x on y** by either using the formula $x = c + dy$

  where $c = \bar{x} - d\bar{y}$ and $d = \dfrac{\sum xy - n\bar{x}\,\bar{y}}{\sum y^2 - n\bar{y}^2}$

  or by using the calculator directly.

- **used the regression line of x on y** to predict x given y (but only over the given range of y, or thereabouts)

- seen that **both regression lines pass through** $(\bar{x}, \bar{y})$

- ignored **outliers** (rogue points) when they clearly don't fit the general scheme of things

- interpreted the **coefficients** α **and** β in the regression line $y = \alpha + \beta x$ as

  $\alpha = $ initial value of y

  $\beta = $ rate at which y grows

- found **best fit curves** by (say) finding the regression line of $\dfrac{y}{x}$ on x.

ANSWERS

Practice questions A

1 (a) $r = 0.901$. Good positive linear correlation.

(b) $r = 1$. Perfect positive linear correlation $(y = 2x - 2)$.

(c) $r = -0.987$. Very good negative linear correlation

(d) $r = -0.472$. Modest negative linear correlation.

2 The scatter diagram suggests good negative linear correlation.

$r = -0.959$, which confirms the above view.

Practice questions B

1 (a) 0.9847
   (b) 0.9905
   (c) 0.9701
   (d) −0.9719

Practice questions C

1 (a) −0.9378
   (b) 0.99497 ...
   (c) 0.9686

2 0.6034 (take h as x and w as y)

Practice questions D

1 (a) $y = 1.95x - 2.55$
   (b) $y = 1.759x - 1.448$

2 $y = 0.652x - 2.594$. Estimate = 36.53, $r = 0.905$

3 $y = 1.49x - 24$. Estimate = 28.4

Practice questions E

1 (a) (i) 0.9705

(ii) $y = 1.168x - 2.59$,

(iii) (7.7, 6.4)
[usually (Shift 1 , Shift 4)]

(b) (i) 0.9516

(ii) $y = 0.154x - 1.89$

(iii) (15.5, 0.5).

Practice questions F

1 $r = 0.850$
$y = 2.1x - 1.9$
(4, 6.5)
14.9
$x = 0.344y + 1.76*$
(4, 6.5) once again!
4.86 (use *)

2 $r = -0.9831$, $y = -1.109x + 13.52$,
$x = -0.8718y + 11.95$, $y = 2.43$ (use y on x),
$x = 1.49$ (use x on y)

Practice questions G

E is the outlier.

$r = -0.823$, $y = -0.628x + 8.189$, $x = -1.079y + 10.579$

Both lines pass through (5.4, 4.8)

When $x = 11 \Rightarrow y \approx 1.3$.
Probably valid since x near to edge of range.

When $y = 16 \Rightarrow x \approx -6.7$.
Invalid since $y = 16$ is well beyond the sample values for y.

Practice questions H

1 When the temperature is zero, the yield is 81.25. For every degree of temperature increase, the yield goes up by 0.62.

2 If a candidate scores zero on Paper I, she should still score 6.05 on Paper II. For every mark scored on Paper I, the mark on Paper II goes up by 0.86.

Practice questions I

1 $\dfrac{y}{x} = 0.45 + 0.295x^2$, 66.42

2 $y = 4.77 - \dfrac{6.9}{x}$, 3.39

7

Discrete random variables

INTRODUCTION   In this section we'll be following up the ideas of sample space and probabilities (Section 5) and linking them with expectation (Section 3) and variance (Section 4). We'll be looking in more depth at the properties of E(X) and Var(X), discovering useful formulae and, finally, we'll apply our skills to a discrete uniform distribution.

By the end of this section you should be able to:

● understand what is meant by a random variable

● understand what is meant by a probability density function and the distribution function

● calculate and interpret an expected value and a variance.

Random variables as models

Random variables provide us with models for data and form the link between probability and statistics. However, one of the major problems of statistics is to provide good models for data which can then be used to make predictions and, possibly, help in formulating general theories about the situation being considered. That gives us the motivation for studying the general properties of random variables.

Discrete random variables

OCR **S1** 5.11.3 (a)

Definition: Suppose X is a variable quantity which takes n discrete values $x_1, x_2, \ldots x_n$ and further that it takes these values with probabilities $p_1, p_2, \ldots p_n$, then if $p_1 + p_2 + \ldots + p_n = 1$, X is called a **discrete random variable**.

At this stage, we must introduce what will strike you as a strange convention.

As in our definition above, we shall use *capital letters* to name random variables – X, Y, Z – but we shall use the corresponding *lower case letters* for particular values of these random variables.

So x or x_i can both be used for particular values of the random variable X.

This explains the strange notation $P(X = x)$ – the probability that the random variable X takes the particular value x.

Example

Consider the experiment of tossing a coin three times and noting the number of tails shown. Show that X – the number of tails shown – is a discrete random variable.

Solution

We have the following outcomes which are mutually exclusive and exhaustive:

Outcome (x)   *Number of tails*   0, 1, 2, 3

We need to show that:

$$\sum_{x=0}^{3} P(X = x) = 1$$

Letting H denote heads and T tails we have for:

$$P(X = 0) \;=\; P(H \cap H \cap H) = \tfrac{1}{2} \times \tfrac{1}{2} \times \tfrac{1}{2} = \tfrac{1}{8}$$

$$P(X = 1) \;=\; P(H \cap H \cap T) + P(H \cap T \cap H) + P(T \cap H \cap H)$$
$$\;=\; (\tfrac{1}{2} \times \tfrac{1}{2} \times \tfrac{1}{2}) + (\tfrac{1}{2} \times \tfrac{1}{2} \times \tfrac{1}{2}) + (\tfrac{1}{2} \times \tfrac{1}{2} \times \tfrac{1}{2}) = \tfrac{3}{8}$$

$$P(X = 2) \;=\; P(H \cap T \cap T) + P(T \cap H \cap T) + P(T \cap T \cap H)$$
$$\;=\; (\tfrac{1}{2} \times \tfrac{1}{2} \times \tfrac{1}{2}) + (\tfrac{1}{2} \times \tfrac{1}{2} \times \tfrac{1}{2}) + (\tfrac{1}{2} \times \tfrac{1}{2} \times \tfrac{1}{2}) = \tfrac{3}{8}$$

$$P(X = 3) \;=\; P(T \cap T \cap T) = \tfrac{1}{2} \times \tfrac{1}{2} \times \tfrac{1}{2} = \tfrac{1}{8}$$

Note that we have to be careful to notice that the same outcome (e.g. two heads) can occur in a number of different ways.

Therefore:

$$\sum_{x=0}^{3} P(X = x) = \tfrac{1}{8} + \tfrac{3}{8} + \tfrac{3}{8} + \tfrac{1}{8} = 1$$

confirming that X is a discrete random variable.

This discrete random variable X, the number of tails when 3 coins are tossed, can be summarised in the following table, called its **probability distribution**.

x	0	1	2	3
$P(X = x)$	$\tfrac{1}{8}$	$\tfrac{3}{8}$	$\tfrac{3}{8}$	$\tfrac{1}{8}$

where the first row lists the outcomes and the second row lists the probabilities for each of the outcomes.

This random variable would be a good model for the experiment providing the coin were unbiased.

The following example illustrates that discrete random variables can be defined by functions.

Example

Consider the function defined as follows:

$$p(x) = kx^2 \qquad x = 1, 2, 3$$
$$p(x) = k(7 - x)^2 \qquad x = 4, 5, 6$$
$$p(x) = 0 \qquad \text{otherwise}$$

Now if $p(x)$ stands for $P(X = x)$ then for a certain value of k, which we can determine, the function will define a discrete random variable.

The problem is to find the value of k.

| Solution | Using the appropriate part of the function we have: |

$$
\begin{aligned}
p(1) &= k \times 1^2 &&= k &&= P(X=1) \\
p(2) &= k \times 2^2 &&= 4k &&= P(X=2) \\
p(3) &= k \times 3^2 &&= 9k &&= P(X=3) \\
p(4) &= k\,(7-4)^2 &&= 9k &&= P(X=4) \\
p(5) &= k\,(7-5)^2 &&= 4k &&= P(X=5) \\
p(6) &= k(7-6)^2 &&= k &&= P(X=6)
\end{aligned}
$$

We require that these probabilities should add up to 1.

$$
\sum_{x=1}^{6} p(x) = 1 \Rightarrow k + 4k + 9k + 9k + 4k + k = 1 \Rightarrow 28k = 1 \Rightarrow k = \frac{1}{28}
$$

We can now complete the probability distribution for the discrete random variable X:

x	1	2	3	4	5	6
$P(X=x)$	$\frac{1}{28}$	$\frac{4}{28}$	$\frac{9}{28}$	$\frac{9}{28}$	$\frac{4}{28}$	$\frac{1}{28}$

Practice questions A

1 A fair dice has three faces marked 1, one face marked 2 and two faces marked 4. Another fair dice has four faces marked 1 and two faces marked 2. These two dice are thrown and the *total score* is recorded. Set up a probability distribution table.

2

Set up a probability distribution table for this spinner.

3 Find the value of k in the following probability distribution.

x	1	2	3	4	5
$P(X=x)$	k	$2k$	$3k$	$2k$	k

4 Write down whether or not there is anything wrong with the following probability distributions. If there is something wrong, explain what it is.

(a)

x	1	2	3
$P(X=x)$	0.2	0.6	0.2

(b) $P(X=x) = \begin{cases} \dfrac{4-x}{5} & x = 1, 2, 3, 4, 5 \\ 0 & \text{otherwise} \end{cases}$

The cumulative distribution function

The cumulative distribution function of a discrete random variable X is defined as:

$$
F(x_0) = P(X \le x_0) = \sum_{x \le x_0} p(x)
$$

and is a new function formed from the probability function by accumulating the probability up to some value x_0 (rather like the cumulative frequencies from Section 2).

Example	The random variable X, is the total score obtained when two dice are thrown. Find:
	(a) The probability distribution of X
	(b) The cumulative distribution function $F(x)$
	(c) Say what is meant by $F(6)$ and $F(12)$

Solution	(a)	The possible outcomes range from $X = 2$ to $X = 12$, but the outcomes are not equally likely as, for example, there are more ways of scoring a total of 6 than a total of 2.

Proceeding systematically the following probabilities are obtained.

x	2	3	4	5	6	7	8	9	10	11	12
$P(X = x)$	$\frac{1}{36}$	$\frac{2}{36}$	$\frac{3}{36}$	$\frac{4}{36}$	$\frac{5}{36}$	$\frac{6}{36}$	$\frac{5}{36}$	$\frac{4}{36}$	$\frac{3}{36}$	$\frac{2}{36}$	$\frac{1}{36}$

where for example the value for $x = 4$ is obtained by counting up the possible ways of obtaining a total of 4 which are (3, 1), (2, 2), (1, 3).

There are 36 possible outcomes altogether. The table is the probability distribution.

(b) For the cumulative distribution function we get:

x_0	2	3	4	5	6	7	8	9	10	11	12
$P(X \le x_0)$	$\frac{1}{36}$	$\frac{3}{36}$	$\frac{6}{36}$	$\frac{10}{36}$	$\frac{15}{36}$	$\frac{21}{36}$	$\frac{26}{36}$	$\frac{30}{36}$	$\frac{33}{36}$	$\frac{35}{36}$	$\frac{36}{36}$

where for example the value for $x_0 = 4$ is obtained by adding all the probabilities up to $x = 4$ in the table from part (a).

so $P(X \le 4) = \frac{1}{36} + \frac{2}{36} + \frac{3}{36} = \frac{6}{36} = F(4)$

(c) $F(6) = \frac{15}{36}$ and equals $P(X \le 6)$

$F(12) = \frac{36}{36}$ and equals $P(X \le 12)$

Note that $F(12) = 1$ and this should be the case as all of the probabilities are added at this point (all scores being ≤ 12 for this random variable).

Practice questions B

1 Find the cumulative distribution function $F(x)$ for question 1 in Practice questions A.

2 Find the cumulative distribution function $F(x)$ for question 2 in Practice questions A.

3 A discrete probability distrubtion is given by:

x	0	1	2	3	4
$P(X = x)$	k	$2k$	$3k$	$4k$	$2k$

(a) Find the value of k

(b) Find the cumulative distribution function $F(x)$.

Expected value

Given the similarities between a frequency distribution and a probability distribution, it may be apparent that there should also be similarities between some of the statistical measures we used in earlier sections on data and the equivalent statistics for a probability distribution. In probability we focus on two such measures: the mean and the variance of the distribution, although the mean is more usually referred to as expectation.

If X is a discrete random variable with probability distribution as in the table shown

x	x_1	x_2	...	x_n
$P(X = x)$	p_1	p_2	...	p_n

then the expectation of X is defined as

$$E(X) = x_1 p_1 + x_2 p_2 + \dots + x_n p_n = \sum_{i=1}^{i=n} x_i p_i$$

An alternative notation for $E(X)$ is μ (pronounced 'mu').

Example

Let X = the outcome when an ordinary die is thrown.

Then X has the distribution shown:

x	1	2	3	4	5	6
$P(X = x)$	$\frac{1}{6}$	$\frac{1}{6}$	$\frac{1}{6}$	$\frac{1}{6}$	$\frac{1}{6}$	$\frac{1}{6}$

$$\therefore \quad E(X) = 1 \times \frac{1}{6} + 2 \times \frac{1}{6} + 3 \times \frac{1}{6} + 4 \times \frac{1}{6} + 5 \times \frac{1}{6} + 6 \times \frac{1}{6} = 3.5$$

If you were to throw a die a large number of times (or simulate this activity using random numbers) and find the mean of your results it would be close to 3.5. The bigger the number of throws or simulations the closer the average result would be to the value of 3.5.

$E(X) = 3.5$ is the theoretical (long-term) mean for the experiment.

Example

(a) A discrete random variable X has the probability function

$$P(X = x) = \begin{cases} kx^2 & x = 0, 1, 2, 3 \\ 0 & \text{otherwise} \end{cases}$$

Find the values of k and $E(X)$

Solution

(a) To find the value of k use the fact that the probabilities must add to 1

$$P(X = 0) = k \times 0^2 = 0 \qquad P(X = 2) = k \times 2^2 = 4k$$
$$P(X = 1) = k \times 1^2 = k \qquad P(X = 3) = k \times 3^2 = 9k$$

Sum $= 14k = 1 \Rightarrow k = \frac{1}{14} \Rightarrow$

x	0	1	2	3
$P(X = x)$	0	$\frac{1}{14}$	$\frac{4}{14}$	$\frac{9}{14}$

$$\therefore \quad E(X) = 0 \times 0 + 1 \times \frac{1}{14} + 2 \times \frac{4}{14} + 3 \times \frac{9}{14} = 2\frac{4}{7}$$

Practice questions C

1 Refer back to Practice questions A.
Find E(X) for questions 1 and 2.

2 Refer back to question 3 in Practice questions B.
Find E(X).

3 Set up a probability distribution table for this spinner.

Find the expected score that results from one spin.

4 The spinner is spun and the score recorded.
However, if a '1' is spun then the spinner is spun
again and the score recorded is *the sum* of the two
spins. The spinner is never spun more than twice.
Set up a probability distribution table for the spinner.
Deduce the mean *total* score obtained.

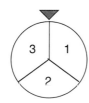

5 A fair dice has faces marked 1, 2, 2, 2, 3 and 4.
Another fair dice has faces marked 1, 3, 3, 3, 3
and 5.

These two dice are thrown and their *total score* is
recorded. Set up a probability distribution table for
the total score.
Deduce the expected total score.

6 A spinner has faces marked 0, 1 and 3. The
probabilities associated with these faces are:

Score	0	1	3
P($X = x$)	0.3	0.5	0.2

Two such spinners are spun and the *total* score
is recorded. Set up a probability distribution table
for the total score and hence find the mean
score.

7 A probability distribution is given as follows:

x	2	3	4	5
P($X = x$)	a	b	0.3	0.2

If E(X) = 3.6, find the values of a and b.

8 Spinners A and B are spun and the *largest score
minus the smallest score* is recorded.

Spinner A	Score	3	4	6
	Prob.	0.4	0.35	0.25

Spinner B	Score	1	2	4
	Prob.	0.5	0.3	0.2

Set up a probability distribution table for the
result and hence find the mean score.

Properties of expectation

OCR **S3** 5.13.3 (a)

The definitions for E(X) extend to E$\big(g(X)\big)$ where g is a function of X and they
are as follows:

For the discrete case with distribution as shown

x	x_1	x_2	...	x_n
P($X = x$)	p_1	p_2	...	p_n

$$E\big(g(x)\big) = g(x_1)\,p_1 + g(x_2)\,p_2 + \ldots + g(x_n)\,p_n = \sum_{i=1}^{n} g(x_i)\,p_i$$

e.g. $E(X^2) = x_1^2\,p_1 + x_2^2\,p_2 + \ldots + x_n^2\,p_n$

| | Example | |

For the discrete distribution:

x	0	1	2	3	4
$P(X = x)$	0.2	0.1	0.1	0.3	0.3

Find:

(a) $E(X)$ (b) $E(X^2)$ (c) $E(5X)$

Solution

(a) $E(X)$ $= \displaystyle\sum_{x=0}^{4} x\ P(X = x)$

$= 0 \times 0.2 + 1 \times 0.1 + 2 \times 0.1 + 3 \times 0.3 + 4 \times 0.3 = 2.4$

(b) $E(X^2)$ $= \displaystyle\sum_{x=0}^{4} x^2\ P(X = x)$

$= 0^2 \times 0.2 + 1^2 \times 0.1 + 2^2 \times 0.1 + 3^2 \times 0.3 + 4^2 \times 0.3 = 8.0$

(c) $E(5X)$ $= \displaystyle\sum_{x=0}^{4} 5x\ P(X = x)$

$= 5 \times 0 \times 0.2 + 5 \times 1 \times 0.1 + 5 \times 2 \times 0.1 + 5 \times 3 \times 0.3 + 5 \times 4 \times 0.3$

$= 12.0$

And so $E(5X) = 5\,E(X)$

This result suggests a number of properties:

> If X is a discrete random variable and a is a constant, then:
>
> (i) $E(a) = a$
>
> (ii) $E(aX + b) = aE(X) + b$
>
> (iii) $E\big(f_1(X) + f_2(X)\big) = E\big(f_1(X)\big) + E\big(f_2(X)\big)$
>
> where f_1 and f_2 are any two functions of X.

Practice questions D

1 If $E(X) = 6$, find the following:

(a) $E(2X)$ (b) $E(5X)$

(c) $E(X + 6)$ (d) $E(2X + 5)$

2 X has a discrete probability distribution given by:

x	0	1	2	3	4
$P(X = x)$	0.2	0.3	0.2	0.2	0.1

Find the following:

(a) $E(X)$ (b) $E(6X)$ (c) $E(X + 6)$

(d) $E(X^2)$ (e) $E(2X^2 + 3)$

3 X has a discrete probability distribution given by:

x	3	4	5
$P(X = x)$	0.6	0.3	0.1

Find the following:

(a) $E(X)$ (b) $E(3X + 2)$

(c) $E(X^2)$ (d) $E(X^2 - 5)$

4 If $E(3X + 5) = 18$ find $E(X)$.

5 If $E(X) = 7$, $E(aX + b) = 23$ and $E(bX + a) = 17$ find the values of a and b.

6 X has a discrete probability distribution given by:

x	2	3	4
$P(X = x)$	0.2	0.4	0.4

Find $E(X^3)$ and hence write down the value of $E(X^3 - 9)$.

The variance of X

OCR **S1** 5.11.3 (a), **S3** 5.13.2 (a)

The definition of variance for a discrete and continous random variables is analogous to the formulae for sample data which we met in Section 4.

It is: $\text{Var}(X) = \sum_{i=1}^{n} (x_i - \mu)^2 \, P(X = x_i)$, where $\mu = E(X)$

It can also be shown that:

$$\text{Var}(X) = E(X^2) - \left(E(X)\right)^2$$

and in practice this is the formula to use.

As with the expected value, there are a number of useful properties.

If X is a random variable and a and b are two constants then:

(a) $\text{Var}(a) = 0$

(b) $\text{Var}(aX) = a^2 \, \text{Var}(X)$

(c) $\text{Var}(aX + b) = a^2 \, \text{Var}(X)$

It is worthwhile investigating these properties in more detail.

Proof of Property (a)

If a is some constant then we have:

$$\text{Var}(a) = E(a^2) - E(a)^2$$

which from the expectation properties we examined earlier implies that:

$$\text{Var}(a) = a^2 - a^2 = 0$$

Intuitively this appears reasonable. Given that the variance can be seen as a measure of dispersion, we know that the dispersion of a constant is zero.

Proof of Property (b)

$$\text{Var}(aX) = E(a^2 X^2) - E(aX)^2 = a^2 E(X^2) - a^2 E(X)^2$$

(from the expectation properties introduced earlier)

$$= a^2 (E(X^2) - E(X)^2) = a^2 \, \text{Var}(X)$$

Proof of Property (c)

$$\text{Var}(aX + b) \quad = a^2 \text{Var}(X)$$

This may seem somewhat surprising until we recollect the meaning of the variance. Effectively we are adding some constant, b, to a variable. This will affect the variable's location (its mean or expected value), but not its dispersion. So two distributions:

$$aX \quad \text{and} \quad aX + b$$

will have the same variance $\text{Var}(aX)$

Consider the probability distribution of the random variable X:

x	0	1	2	3	4
$P(X = x)$	0.2	0.1	0.1	0.2	0.4

Find $E(X)$ and $\text{Var}(X)$ for this distribution.

$$E(X) \quad = \Sigma x\, P(X = x) = 2.5$$

$$E(X^2) \quad = \Sigma x^2 P(X = x) = 0^2 \times 0.2 + 1^2 \times 0.1 + 2^2 \times 0.1 + 3^2 \times 0.2 + 4^2 \times 0.4$$

$$= 8.7$$

$$\text{Var}(X) \quad = E(X^2) - E(X)^2 = 8.7 - 2.5^2 = 2.45$$

Practice questions E

1 Use your calculator to find Var(X) for questions 3 to 8 in Practice questions C.

2 If $E(X) = 5$ and $E(X^2) = 90$ find Var(X).

3 If Var(X) = 30 and $E(X) = 4$ find $E(X^2)$.

4 If Var(X) = 12 and $E(X^2) = 76$ find $E(X)$.

5 If Var(X) = 16 find the following:
(a) Var($2X$)   (b) Var($3X$)
(c) Var($X + 5$)   (d) Var($2X + 5$)

6 If Var(X) = 6 find the following:
(a) Var($3X$)   (b) Var($X + 3$)
(c) Var($X - 3$)   (d) Var($5 + 2X$)

7 If $E(X) = 5$, Var(X) = 6, $E(aX + b) = 11$ and Var($aX + b$) = 54, find the values of a and b, when a is positive.

The uniform distribution

A random variable follows a **discrete uniform distribution** if X is discrete and each value of X has the same probability of occurring. More formally we can state that:

$$P(X = x_i) \quad = \frac{1}{n} \quad x = 1, 2, 3 \ldots n$$

$$= 0, \text{ otherwise}$$

A simple example of such a distribution would relate to the throwing of a six-sided die. Each side (each number from 1 to 6) has the same probability of appearing, that is, $\frac{1}{6}$.

Example

A bank clerk has noted that at a particular period during the day he can deal with between one and five customers in a given ten-minute period (the exact number of course will depend on what each customer wants to do). He has also noted that the number of customers dealt with follows a uniform distribution.

(a) Calculate the probability that in a ten-minute period the clerk will deal with at least four customers.

(b) Calculate the mean and variance of the distribution.

Solution

We have a uniform distribution such that:

Number of customers, x	1	2	3	4	5
$P(X = x)$	0.2	0.2	0.2	0.2	0.2

For part (a), at least four customers implies four or five customers, with a probability:

$P(4 \text{ or } 5 \text{ customers}) = 0.2 + 0.2 = 0.4$

(b) The mean of the distribution is calculated from the appropriate formula:

$E(X) = \Sigma\, xP(X = x) = 1 \times 0.2 + 2 \times 0.2 + 3 \times 0.2 + 4 \times 0.2 + 5 \times 0.2 = 3$

That is, on average three customers will be dealt with by the clerk (hardly a surprising result given the symmetrical distribution).

The variance is given by:

$Var(X) = E(X^2) - E(X)^2 = \Sigma\, x^2\, P(X = x) - 3^2 = 11 - 9 = 12.$

It can be shown that:

if X has a discrete uniform distribution then

- $P(x = x) = \dfrac{1}{n}$ for $x = 1, 2, \ldots n$

  $= 0$, otherwise

- $E(X) = \dfrac{n + 1}{2}$

- $Var(X) = \dfrac{n^2 - 1}{12}$

Our previous example confirmed this for

$n = 5 \Rightarrow E(X) = \dfrac{5H}{2} = 3$

and $n = 5 \Rightarrow Var(X) = \dfrac{5^2 - 1}{12} = 2$

Practice questions F

1 X takes the values 1, 2, 3, ... 7 and has a discrete uniform distribution. Set up a probability distribution table for X and write down the values of $E(X)$ and $Var(X)$.

2 X takes the values 1, 2, 3, ... 12 and has a discrete uniform distribution. Find $P(X > 8)$ and write down the values of $E(X)$ and $Var(X)$.

3 X has a discrete uniform distribution and $E(X) = 11$. Find $Var(X)$.

4 X has a discrete uniform distribution and $Var(X) = 2$. Find $E(X)$.

5 (a) Can X have a discrete uniform distribution with mean 7.5?

   (b) Can X have a discrete uniform distribution with mean $6\frac{1}{3}$?

   (c) Can X have a discrete uniform distribution with variance 9?

6 X has a discrete uniform distribution and $E(X) = 12$. Find $P(X < 7)$ and $Var(X)$.

SUMMARY EXERCISE

1 X is the random variable 'The number of red balls selected when 3 balls are drawn at random' from a bag which contains 5 red, 4 blue and 1 white ball.

   Work out the probability distribution of X and find $P(X > 1)$.

2 In a game of chance, 3 fair coins are tossed. The score, X, is defined as the number of heads showing. Find the probabilities of every possible value of X and hence show that X is a discrete random variable.

3 The random variable X has the distribution given by the table

r	1	2	3	4
$P(X = r)$	k	$\frac{k}{2}$	$\frac{l}{3}$	$\frac{l}{4}$

In addition it is known that:

$P(X \leq 2) = 2\,P(X > 2)$

Find $P(X = 2)$

4 Find the expectation and variance of the PDF you have found in Exercise 2.

5 A discrete random variable X has a probability function given by:

$$p(x) = kx^2 \quad\left.\vphantom{\begin{matrix}a\\b\end{matrix}}\right\} \quad x = 1, 2, 3$$
$$p(x) = 0 \quad\qquad\qquad \text{otherwise}$$

Find the value of:

   (a) k (b) $E(X)$ (c) $Var(X)$.

6 If $E(X) = 9$ find the values of $E(2X)$, $E(X - 2)$ and $E(2X + 5)$.

7 If $Var(X) = 5$ find the values of $Var(2X)$, $Var(X - 2)$ and $Var(2X + 5)$.

8 X has a discrete uniform distribution and $E(X) = 14$. Find $P(X > 20)$ and $Var(X)$.

SUMMARY

In this section we have seen that if X has a **discrete probability distribution** given by:

x	x_1	x_2	...	x_n
$P(X = x)$	p_1	p_2	...	p_n

then:

- $p_1 + p_2 + ... + p_n = 1$
- the **cumulative function** $F(x)$ is defined by $F(a) = p_1 + p_2 + ... + p_a$
- the **expected value** $E(X)$, where $E(X) = x_1 p_1 + x_2 p_2 + ... + x_n p_n$, has the properties:

    $E(a) = a$

    $E(aX + b) = a\,E(X) + b$, with a and b constants.

- the **variance** $Var(X)$, where $Var(X) = E(X^2) - [E(X)]^2$, has the properties:

    $Var(a) = 0$

    $Var(aX) = a^2 Var X$

    $Var(aX + b) = a^2 Var X$, with a and b constants

We have also seen that a **discrete uniform distribution** has:

- $P(X = x) = \dfrac{1}{n}$ for $x = 1, 2, ... n$

    $= 0$, otherwise
- $E(X) = \dfrac{n + 1}{2}$ and $Var(X) = \dfrac{n^2 - 1}{12}$

ANSWERS

Practice questions A

1

x	2	3	4	5	6
$P(X = x)$	$\frac{1}{3}$	$\frac{5}{18}$	$\frac{1}{18}$	$\frac{2}{9}$	$\frac{1}{9}$

(see sample space example in Section 5, page 62)

2

x	1	2	3
$P(X = x)$	$\frac{1}{4}$	$\frac{1}{2}$	$\frac{1}{4}$

3 $k = \dfrac{1}{9}$

4 (a) OK. Probabilities add to 1

  (b) No good. $P(X = 5)$ is negative.

Practice questions B

1

x	2	3	4	5	6
$F(x)$	$\frac{1}{3}$	$\frac{11}{18}$	$\frac{2}{3}$	$\frac{8}{9}$	1

2

x	1	2	3
$F(x)$	$\frac{1}{4}$	$\frac{3}{4}$	1

3 $k = \dfrac{1}{12}$

x	0	1	2	3	4
$F(x)$	$\frac{1}{12}$	$\frac{1}{4}$	$\frac{1}{2}$	$\frac{5}{6}$	1

Practice questions C

1   E(X) = 3.5 and E(X) = 2

2   E(X) = $2\frac{1}{3}$

3

x	1	2	3	4	5
P(X = x)	$\frac{1}{8}$	$\frac{1}{4}$	$\frac{1}{4}$	$\frac{1}{4}$	$\frac{1}{8}$

∴ E(X) = 3

4

x	2	3	4
P(X = x)	$\frac{4}{9}$	$\frac{4}{9}$	$\frac{1}{9}$

∴ E(X) = $2\frac{2}{3}$

5

x	2	3	4	5	6	7	8	9
P(X = x)	$\frac{1}{36}$	$\frac{3}{36}$	$\frac{5}{36}$	$\frac{13}{36}$	$\frac{5}{36}$	$\frac{7}{36}$	$\frac{1}{36}$	$\frac{1}{36}$

∴ E(X) = $5\frac{1}{3}$

6

x	0	1	2	3	4	6
P(X = x)	0.09	0.3	0.25	0.12	0.2	0.04

∴ E(X) = $2\frac{1}{5}$

7   a = 0.1 and b = 0.4

[Hint: 2a + 3b + 1.2 + 1 = 3.6 and
a + b + 0.3 + 0.2 = 1]

8

x	0	1	2	3	4	5
P(X = x)	0.07	0.2	0.355	0.175	0.075	0.125

∴ E(X) = 2.36

Practice questions D

1   (a) 12   (b) 30   (c) 12   (d) 17

2   (a) 1.7   (b) 10.2   (c) 7.7   (d) 4.5   (e) 12

3   (a) 3.5   (b) 12.5   (c) 12.7   (d) 7.7

4 $4\frac{1}{3}$

5   a = 3, b = 2

[Hint: 7a + b = 23 and a + 7b = 17]

6   38, 29

Practice questions E

1   (a) 1.5   (b) $\frac{4}{9}$ (c) $2\frac{2}{9}$

(d) 2.18   (e) 0.84   (f) $1\frac{594}{625}$

2   65

3   46

4   8

5   (a) 64   (b) 144   (c) 16   (d) 64

6   (a) 54   (b) 6   (c) 6   (d) 24

7   a = 3, b = −4

Practice questions F

1

x	1	2	3	4	5	6	7	E(X) = 4
P(X = x)	$\frac{1}{7}$	$\frac{1}{7}$	$\frac{1}{7}$	$\frac{1}{7}$	$\frac{1}{7}$	$\frac{1}{7}$	$\frac{1}{7}$	Var(X) = 4

2 $\frac{4}{12} = \frac{1}{3}$, E(X) = 6.5, VAR(X) = $11\frac{11}{12}$

3 $36\frac{2}{3}$

4   3 (because n = 5)

5   (a) yes, n = 14

(b) No, n = $11\frac{2}{3}$!

(c) No, n = $\sqrt{109}$!

6 $\frac{6}{23}$, 44 (because n = 23).

8

The normal distribution

In the previous section we studied discrete distributions. In this section we are going to look at an important *continuous* distribution – *the normal distribution.*

The normal distribution is important because many distributions that occur in real life can be successfully modelled by one. For example, heights of females, weights of students, the life times of light bulbs, the breaking strain of car-door catches and the weights of packets of sugar can all be successfully modelled by a normal distribution.

Even distributions which aren't normal can be regarded as being approximately normal, in certain circumstances. For example, if I throw 100 coins and count the number of heads, the distribution obtained (although discrete) can be satisfactorily modelled by a normal distribution with mean 50 and standard deviation 5.

And even distributions which are nowhere near normal (the number of goals scored by your favourite football team, for example) produce a normal model if we take sufficiently large samples and compute the mean number of goals per sample.

Then there is quality control in industry ... yes, you've guessed it, the normal distribution is used.

The normal distribution then is the most important in statistics because it provides a suitable model for a wide range of data.

The normal distribution

OCR **S2** 5.12.2 (a)

If X is a random variable having a normal distribution, then we write $X \sim N(\mu, \sigma^2)$.

The parameters of the distribution are actually the mean and variance of the distribution, i.e.

$E(X) = \mu$ and $Var(X) = \sigma^2$

A variable which is normally distributed will typically exhibit a characteristic 'bell-shape' as illustrated in Fig. 8.1.

Figure 8.1

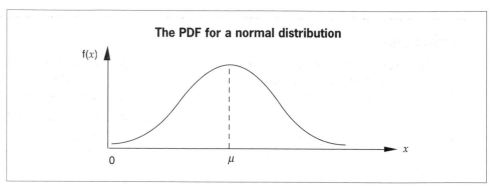

The PDF for a normal distribution

It can be seen that the distribution is symmetrical about the mean value, μ. It is evident that the curve itself will alter depending on the two parameters μ and σ^2. It is worthwhile considering what these alterations would be.

Example

(a) Assume we have a number of normally distributed variables with the same variance but differing mean values. Draw a sketch of the distributions.

(b) Now assume these variables have the same mean but differing variances. Sketch these distributions now.

Solution

For case (a) we would expect a diagram similar to Fig. 8.2, whilst for case (b), it would be similar to Fig. 8.3.

Figure 8.2

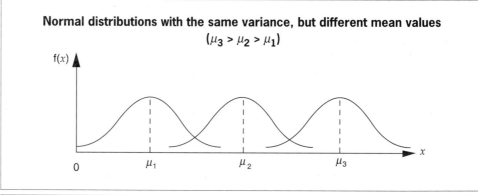

Normal distributions with the same variance, but different mean values
$(\mu_3 > \mu_2 > \mu_1)$

Figure 8.3

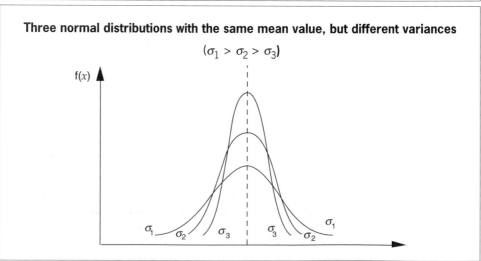

Three normal distributions with the same mean value, but different variances
$(\sigma_1 > \sigma_2 > \sigma_3)$

Practice questions A

1 The distributions below both have a mean of 50. However, one has a standard deviation of 3 and the other of 5. Which is which?

(a)   (b)

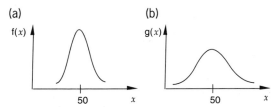

2 The time taken to run 100m is shown in the distribution below. The mean is 10.4 sec and the standard deviation is 0.5 sec. Mark in the mean and shade the area within one standard deviation of the mean.

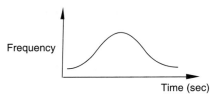

3 The distribution below represents the weights of a population of snails. If the mean weight is 15g and the standard deviation is 3g:

(a) mark in 15g on the weight axis

(b) shade the area that represents more than 2 standard deviations away from the mean.

4 If $X \sim N(10, 30)$ write down the values of $E(X)$ and $Var(X)$.

5 If $X \sim N(12, 25)$, what is the standard deviation of X?

Calculation of probabilities

OCR S2 5.12.2 (b)

Let's start with an example.

Example	Suppose we have a random variable $X \sim N(100, 25)$, i.e. X has a normal distribution with mean = 100 and variance = 25.

Suppose further that we need to find $P(X > 110)$.

Then we would need to find the area illustrated in Fig. 8.4.

Figure 8.4	

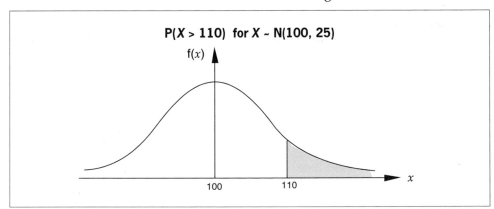

The probabilities for all normal distributions are obtained from tables of the standardised normal distribution.

The standardised normal distribution

The standardised normal distribution takes advantage of the fact that all normal distributions have the same basic shape, but vary only in terms of two parameters: mean and variance. This is effectively the same as saying that all such distributions are identical except for the values of μ and σ^2.

In general if $X \sim N(\mu, \sigma^2)$

and we let $Z = \dfrac{X - \mu}{\sigma}$ then Z is $\sim N(0, 1)$.

Every normal distribution can be transformed into a standard normal in this way. We write $\Phi(z)$ for $P(Z \le z)$ and it is values of $\Phi(z)$ which can be found in the table in Appendix 2 (p. 138). Φ is the Greek capital letter 'phi'.

We will now see how to calculate normal probabilities by an example.

Example	If $X \sim N(100, 25)$, find:

(a) $P(X > 110)$

(b) $P(X < 95)$

(c) $P(95 < X < 110)$

Solution	The first step *always* when finding normal probabilities is to standardise the random variable.

Let $\qquad Z = \dfrac{X - 100}{5}$

(Note that the denominator of this fraction is σ whereas the parameter of the distribution is σ^2.)

(a) $P(X > 110) = P\left(Z > \dfrac{110 - 100}{5} \right) = P(Z > 2)$

We have transformed the problem from

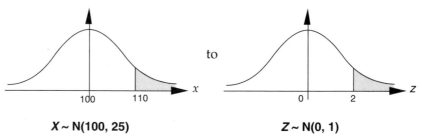

to

X ~ N(100, 25)   **Z ~ N(0, 1)**

The diagram illustrates the fact that $P(X > 110) = P(Z > 2)$

Now we need the tables in the Appendix.

The left-hand column gives values for Z. It is important to notice that the table gives probabilities of the form $P(Z < z)$ where z is a positive value. In order to calculate probabilities not of this form we use the symmetry of the curve and the fact that the total area under the curve is 1 as appropriate.

For $P(Z > 2)$ we use the fact that the total area under the curve is 1 and find $1 - P(Z \le 2)$, or equivalently $P(Z < 2)$, since $P(Z = 2) = 0$ for a continuous distribution:

i.e. $P(Z > 2) = 1 - P(Z \le 2) = 1 - 0.9772 = 0.0228$

Hence $P(X > 110) = 0.023$ (3 d.p.)

(b) $P(X < 95) = P\left(Z < \dfrac{95 - 100}{5}\right) = P(Z < -1)$

Now for $P(Z < -1)$ we cannot use the tables directly, but by symmetry

$P(Z < -1) = P(Z > 1)$

The problem has been transformed from:

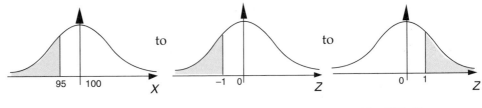

X ~ N(100, 25)   **Z ~ N(0, 1)**   **Z ~ N(0, 1)**

So $P(Z < -1) = 1 - P(Z \le 1)$

(using the fact that the total area is 1)

$$= 1 - 0.8413 = 0.1587$$

Hence $P(X < 95) = 0.159$ (3 d.p.)

(c) $P(95 < X < 110) = P\left(\dfrac{95 - 100}{5} < Z < \dfrac{110 - 100}{5}\right) = P(-1 < Z < 2)$

So now we have to find:

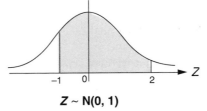

Z ~ N(0, 1)

Clearly from our answers to (a) and (b) this will be

$$1 - 0.023 - 0.159 = 0.818$$

i.e. $P(95 < X < 110) = 0.82$ (2 d.p.)

It is important to be able to use normal distribution tables efficiently and correctly. It is also worth noting that a simple sketch of the areas being found is a useful device and can help decide whether answers obtained are sensible.

| **Example** | In an examination it is known that the distribution of marks is N(52, 24). |

(a)   What proportion of marks will exceed 55?

(b)   What proportion of marks will be less than 45?

(c) If Grade A is to be awarded to the top 5% of marks, what mark must someone achieve to get this grade?

(d) The bottom 20% of marks are classed as Grade F. What range of marks does this represent?

Solution

(a) For marks greater than 55:

$$X > 55 \implies Z > \frac{55 - 52}{4.9} = 0.61$$

giving a probability value of $P(Z < 0.61) = 0.7291$. But we require $P(Z > 0.61)$ hence we have a probability of $(1 - 0.7291) = 0.2709$. That is, just over 27% of marks will exceed 55.

(b) For marks less than 45 we have $Z \leq -1.43$

By symmetry, $P(Z \leq -1.43) \qquad = P(Z \geq 1.43)$ (see diagrams below)

$$= 1 - 0.9236 = 0.0764 \text{ or almost 8\%.}$$

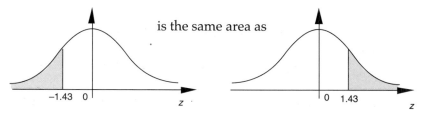

(c) To find the top 5% mark requires a different approach. This time we know the proportion (5% or 0.05, corresponding to a probability of 0.95) but require the x-value that corresponds to this. Searching through the table we see that the z value corresponding to 0.95 is 1.645 (given that z values of 1.64 and 1.65 are both equidistant). So we have:

$$z = 1.645 = \frac{x - 52}{4.9} \text{ which gives } x = 60.06$$

In other words a mark of 60 or more will be achieved by 5% of people taking the exam.

(d) A similar approach can now be adopted to find the bottom 20% mark. For 20% (or 0.20) the nearest corresponding z value is 0.84. But remembering which side of the distribution we are looking at this implies:

$$z = -0.84 = \frac{x - 52}{4.9} \text{ which gives } x = 47.88$$

If we rounded this to give a practical result we would say a mark below 48 would be graded as F.

Let's now look at a normal distribution problem that involves some algebra. In particular, a question in which the mean is unknown.

Example	A machine produces components to any required length specification with a standard deviation of 1.5 mm. If all lengths less than 90 mm are to be rejected, and if this rejection rate is to be 2.5%, to what value should the mean length be adjusted? (Assume the distribution of lengths to be normal.)

Solution	The question leads us to the following sketch:

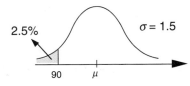

Looking up the 'long tail' of 97.5% in the normal tables, we therefore get:

$$\frac{90 - \mu}{1.5} = -1.96$$

$\therefore \quad 90 - \mu = -2.94$

$\therefore \quad \mu = 92.94$

$\therefore$ The mean length should be adjusted to 92.94 mm.

Finally, let's look at an example that leads to simultaneous equations.

Example	Observation of a very large number of cars at a certain point on a motorway establishes that the speeds are normally distributed. 97.5% of cars have speeds less than 130 km/hr, and 33% of cars have speeds less than 110 km/hr. Determine the mean speed μ and the standard deviation σ.

Solution	The question leads us to the following sketch:

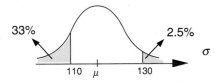

Looking up the 'long tail' of 97.5% in the normal tables, we therefore get:

$$\frac{130 - \mu}{\sigma} = 1.96, \quad \text{i.e. } 130 - \mu = 1.96\sigma \qquad [1]$$

And looking up the 'long tail' of 67%, we get:

$$\frac{110 - \mu}{\sigma} = -0.44 \quad \text{i.e. } 110 - \mu = -0.44\,\sigma \qquad [2]$$

The quickest way of solving the simultaneous equations [1] and [2] is to *subtract one from the other*:

$\therefore \quad 20 = 2.4\sigma \qquad \therefore \sigma = 8\frac{1}{3}$

Substituting this value for σ in [1] gives us

$$130 - \mu = 16\frac{1}{3} \quad \therefore \mu = 113\frac{2}{3}$$

$\therefore$ The mean speed is $113\frac{2}{3}$ km/hr with a standard deviation of $8\frac{1}{3}$ km/hr.

Practice questions B

1 If $Z \sim N(0, 1)$ find:
   (a) $P(0 \le Z \le 1.4)$
   (b) $P(Z \ge 1.7)$
   (c) $P(Z \le -0.6)$
   (d) $P(-1 \le Z \le 2)$
   (e) $P(1 \le Z \le 2)$

2 (a) If $X \sim N(80, 9)$ find $P(77 \le X \le 82)$
   (b) If $X \sim N(50, 16)$ find $P(52 \le X \le 54)$
   (c) If $X \sim N(20, 9)$ find $P(X \le 16)$
   (d) If $X \sim N(50, 64)$ find $P(X \ge 53)$

3 If $Z \sim N(0, 1)$ find:
   (a) a if $P(0 \le Z \le a) = 0.17$
   (b) b if $P(b \le Z \le 0) = 0.17$
   (c) c if $P(Z \ge c) = 0.12$

4 (a) If $X \sim N(80, 16)$ find x if $P(X \ge x) = 0.1$
   (b) If $X \sim N(80, 16)$ find x if $P(X \le x) = 0.05$
   (c) If $X \sim N(12, 4)$ find x if $P(X \ge x) = 0.88$

5 (The following diagrams represent normal distributions.)

   (a)

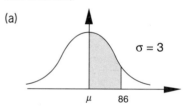

   Shaded area = 0.38. Find μ.

   (b)

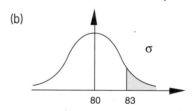

   Shaded area = 0.12. Find σ.

6 (The following diagram represents a normal distribution.)

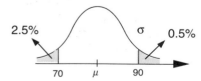

   (2.5% are less than 70 and 0.5% exceed 90.)

   Use simultaneous equations to find μ and σ.

7 A product is sold in packets marked 500g. The mean weight is in fact 510g. Assuming a normal distribution for the weights find:
   (a) the percentage underweight if the standard deviation is 4g
   (b) the standard deviation if 1% of the packets are underweight.

8 The speeds of cars passing a certain point on a motorway can be taken to be normally distributed. Observations show that of cars passing the point, 95% are travelling at less than 85 mph and 10% are travelling at less than 55 mph.
   (a) Find the average speed of the cars passing the point.
   (b) Find the proportion of cars that travel at more than 70 mph.

9 Tests on a certain brand of electric lamp indicate that the length of life is $N(1700, 400)$ measured in hours. Estimate the percentage of lamps which can be expected to burn:
   (a) more than 1725 hours
   (b) less than 1690 hours
   (c) between 1680 and 1710 hours.

10 In a particular school 1460 pupils were present on a particular day. By 8.40 a.m. 80 pupils had already arrived, and at 9.00 a.m. 12 pupils had not arrived but were on their way to school. Assuming that arrival times are normally distributed find:
   (a) the time by which half of those eventually present had arrived
   (b) the standard deviation of the times of arrival.
   If registration occurred at 8.55 a.m. how many would not have arrived by then?

   If each school entrance permitted a maximum of 30 pupils per minute to enter, find the maximum number of entrances required to cope with the 'peak' minute of arrival.

SUMMARY EXERCISE

1 Find the shaded areas in the following normal distributions:

(a)

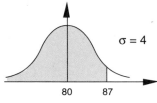

(b)

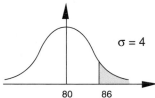

(c)

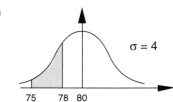

(d)

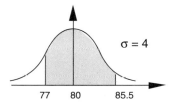

(e)

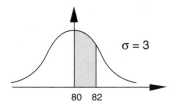

(f)

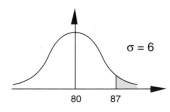

2 Find the value of x in each of the following normal distributions:

(a)

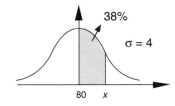

(b)

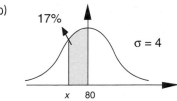

(c)

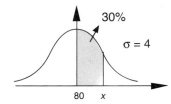

(d)

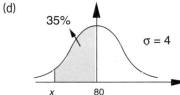

3 (a) $X \sim N(\mu, 25)$ and $P(X < 32) = 0.6$. Find μ.

(b) $X \sim N(40, \sigma^2)$ and $P(X > 37) = 0.7$. Find σ.

(c) $X \sim N(\mu, \sigma^2)$ and it is known that:
$P(X < 22) = 0.74$
$P(X > 17) = 0.78$
Find μ and σ^2.

4 It is found that 9.68% of a certain population of normally distributed length are under 40 cm and 1.5% are over 48 cm. Calculate the mean and variance of the distribution.

5 An automatic filling machine is known to operate with a standard deviation of 1.5g. To what 'average filling' should the machine be set so that 95% of the packets are over 250g?

SUMMARY   In this section we have seen that:

- a **normal distribution** is represented by a bell-shaped curve

- $X \sim N(\mu, \sigma^2)$ means that X is normally distributed with **mean** μ and **variance** σ^2

- if we let $Z = \dfrac{x - \mu}{\sigma}$, then $Z \sim N(0, 1)$

- the standardised normal curve $Z \sim N(0, 1)$ is symmetrical about $z = 0$ and has area equal to 1

- **simultaneous equations** are required if we have to find both μ and σ.

ANSWERS

Practice questions A

1   (a) has a standard deviation of 3

2

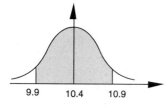

9.9   10.4   10.9

3

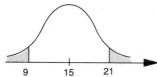

9   15   21

4 $E(X) = 10$, $Var(X) = 30$.

5 $\sqrt{25} = 5$

Practice questions B

1   (a) 0.419   (b) 0.0446
        (c) 0.274   (d) 0.8185
        (e) 0.136

2   (a) 0.589   (b) 0.150
        (c) 0.0912   (d) 0.354

3   (a) 0.44   (b) −0.44   (c) 1.175

4   (a) 85.1   (b) 73.4   (c) 9.65

5   (a) 82.5   (b) 2.55

6 $\mu = 78.6$, $\sigma = 4.41$

7   (a) 0.62%   (b) 4.29g

8   (a) 68.1 mph (b) 0.428

9   (a) 10.6%   (b) 30.85%   (c) 53.3%

10 (a) 8.48 a.m.   (b) 5 mins; 118; 4 entrances

$S1$

Practice examination paper

(Attempt all 9 questions.)

1 Explain what you understand by:

   (a) a population   (b) a sample.

2 Study the following stem and leaf diagram.

4	0 1
4	5 5 7
5	0 0 3
5	5 5 6 8
6	0 1 2 2 3
6	5 5 6 7 7 8 9 9
7	0 1 3 3 4 4 4
7	6 7 7 8 8 8 9

   (4|0 means 4·0, 4|5 means 4·5, and so on.)

   (a) How many readings are shown?
   (b) Find the median and quartiles of the data.
   (c) Represent the data with a box and whisker diagram.
   (d) Calculate the mean and standard deviation for the data.
   (e) Comment on the skewness of the distribution.

3 Four sets of bivariate data P, Q, R and S, are displayed in scatter diagrams. State which of P, Q, R and S have product moment correlation coefficients close to –1.

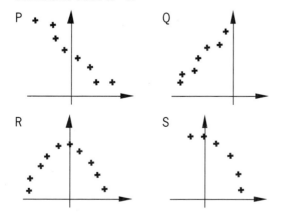

4 A discrete random variable X has a probability distribution as shown below.

x	3	4	5	6	7	8
$P(X = x)$	0.1	0.2	0.35	0.15	0.15	0.05

   Find:

   (a) $P(X < 6)$ (b) $F(5.4)$ (c) $E(X)$
   (d) $Var(X)$ (e) $E(X^2)$ (f) $E(2X^2 + X - 5)$
   (g) $Var(2X - 5)$.

5 Two firms, Glow and Beam, produce light bulbs.

   (a) The lifetimes of Glow's output may be modelled by a normal distribution with mean 450 hrs and standard deviation 15 hrs. What is the probability that the lifetime of one lightbulb selected at random from Glow's output is between 430 and 480 hrs?

   (b) It is found that, for Beam's output, 5% of lifetimes are less than 430 hrs and 12% of lifetimes are more than 455 hrs. Again assuming a normal distribution, calculate the mean and standard deviation of Beam's lifetimes.

6 The letters of the word PERMUTATION are written, one on each of 11 separate cards. The cards are laid out in a line.

   (a) Calculate the number of different arrangements of these letters.

   (b) Find the probability that the vowels (A E I O U) are placed together.

7 The medical test for a certain infection is not completely reliable: if an individual has the infection there is a probability of 0.95 that the test will prove positive, and if an individual does not have the infection there is a probability of 0.1 that the test will prove positive. In a certain population, the probability that an individual chosen at random will have the infection is 0.2.

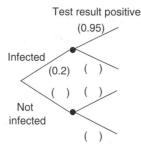

Test result positive
(0.95)

Infected
(0.2)   (   )

(   )   (   )

Not infected

(   )

(a) Complete the tree diagram above to show this information.

(b) Calculate the probability that, if an individual is chosen at random and tested, the result will be negative.

(c) Suppose that 10,000 randomly chosen individuals are tested. Show that 2,700 altogether will be expected to have a positive result. Calculate an estimate of the proportion of these individuals with positive test results that are actually infected.

8 The three events A, B, C are defined in the same sample space.

(a) The events A and B are independent. Interpret this result in terms of probabilities.

(b) The events A and C are mutually exclusive. Interpret this result in terms of probabilities.

Furthermore it is known that

$P(A) = \frac{3}{4}$, $P(C) = \frac{1}{5}$ and $P(A \cup B) = \frac{7}{8}$.

(c) Find:
   (i) $P(A \cup C)$
   (ii) $P(B)$

It is also known that $P(B \cup C) = \frac{13}{20}$.

(d) Find $P(B \mid C)$

9 Study the bivariable data below.

x	2	4	7	11	9	4	13	11	3	10	5
y	3	5.2	8.2	6.9	8	5.4	5.1	6.1	4.1	7.4	6.7

(a) Calculate the product moment correlation coefficient. Comment on its value.

(b) Represent the above data with a scatter diagram and hence comment on your result in (a).

(c) Draw the line $x = 8$ on your scatter diagram. Find:
   (i) the product moment correlation coefficient
   (ii) the regression line of y on x in the cases
      (a) $x < 8$ (b) $x > 8$

Draw in the two regression lines on your scatter diagram.

(d) Use your regression lines from (c) to estimate:
   (i) y when $x = 6$
   (ii) y when $x = 12$
   (iii) y when $x = 8$.
   Discuss this answer to (iii).

(e) Suppose that the original data of 11 pairs of results came from an experiment. Suppose further that a 12th reading for y was 4.8 but the corresponding value for x had been lost. By considering suitable regression lines for $x < 8$ and $x > 8$, find two possible estimates for the missing value of x.

S1
Solutions

Section 1

1 There are a number of reasons that spring to mind. Such a data collection exercise would:

- take too much time
- be very difficult to organise
- be very costly
- provide too much data for us to analyse
- probably be out of date by the time it was completed.

2 (a) (i) The population may be very large and/or inaccessible.

   Costs are kept lower.

   Results are more quickly obtained.

   The testing might destroy the sample.

   (ii) A sampling frame is a list of all the members of a population.

   (b) A list of all the pupils in the school.

3 (a) continuous (d) discrete
   (b) qualitative   (e) continuous
   (c) discrete   (f) continuous

Section 2

1 A frequency table for females would look like this:

Heights (cm)	Number of females
150–	2
155–	7
160–	10
165–	13
170–	15
175–	2
180–	1
185–	0
Total frequency	50

2 Below is a frequency table for females showing relative and cumulative frequencies:

Heights (cm)	No. of females	Relative frequency	Cumulative frequency
150–	2	0.04	2
155–	7	0.14	9
160–	10	0.20	19
165–	13	0.26	32
170–	15	0.30	47
175–	2	0.04	49
180–	1	0.02	50
185–	0	0.00	50
Total	50	1.00	50

3 You should have a diagram similar to this:

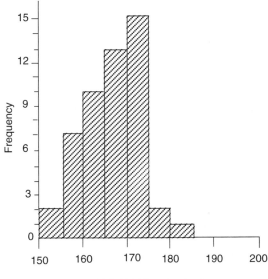

Frequency histogram of female heights

Height of females (cm)

4 The stem and leaf diagrams look like this:

(a)

0	2 6
1	2 2 5 8
2	0 0 7
3	1 8
4	2

(b)

0	5 9
1	0 2 5 6 8
2	1 2 3 5 8 9
3	0 2 3 4 7
4	2 8

5 For females:

lowest value	≈ 150 cm
highest value	≈ 185 cm
lower quartile	≈ 162 cm
upper quartile	≈ 172 cm
median	≈ 168 cm

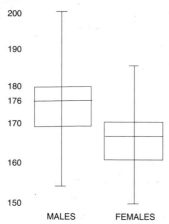

MALES   FEMALES

This clearly shows that females are on average shorter than males. The middle 50% of females are approximately 10 cm shorter than the middle 50% of males.

6 Since the data is continuous, it is really as in the following table.

x	f	Length of interval	Frequency density
9.5–19.5	20	10	2
19.5–24.5	20	5	4
24.5–29.5	15	5	3
29.5–30.5	14	1	14
30.5–34.5	16	4	4
34.5–39.5	10	5	2
39.5–59.5	10	20	0.5

Histogram showing time to answer telephone calls to switchboard

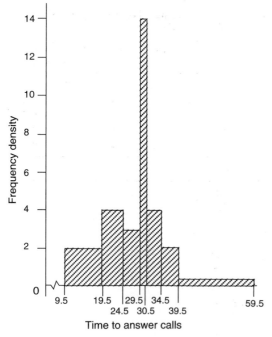

Time to answer calls

The reason for using a histogram is that the data is continuous.

7 No vertical scale given.

No adjustment in heights of bars for differing widths of intervals.

$$\text{Median} = 5 + \frac{13}{65} \times 5 \approx 6.0$$

The number experiencing delays of less than 4.7 minutes

$$= 35 + 34 + 50 + \left(\frac{1.7}{2} \times 36\right) = 149.6 = 50\%$$

8 The frequency table is shown below:

Class	Class width	Frequency	Frequency density
0–4	5	4462	892.4
5–15	11	12,214	1110.4
16–24	9	10,898	1210.9
25–44	20	19,309	965.5
45–74	30	22,820	760.7
75–95	21	3364	160.2

This gives the following histogram.

Histogram showing age distribution of Copeland, Cumbria

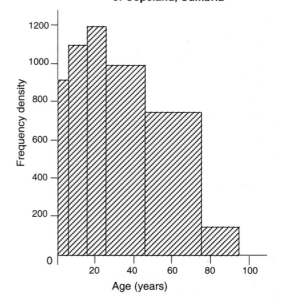

Age (years)

Section 3

1 The appropriate calculations for females summarise to:

Mean $= \dfrac{\Sigma fx}{\Sigma f} = \dfrac{8335.0}{50} = 166.7$ cm

The reason why the mean based on aggregated data is different from that based on raw data is explained on page 30.

2 Mean $= \dfrac{\Sigma x}{n} = \dfrac{54}{9} = 6$

Median $= \dfrac{9+1}{2} = $ 5th value

In increasing order the set becomes

$$2 \quad 3 \quad 4 \quad 6 \quad \underline{7} \quad 7 \quad 8 \quad 8 \quad 9$$

∴ Median = 7

3 Mean $= \dfrac{\Sigma x}{n} = \dfrac{42}{6} = 7$

Median $= \dfrac{6+1}{2} = $ 3.5th value

In increasing order the set becomes

$$1 \quad 4 \quad 6 \quad 8 \quad 11 \quad 12$$

∴ Median $= \dfrac{6+8}{2} = 7$

4 $\bar{x} =$

$\dfrac{5 \times 2 + 6 \times 7 + 7 \times 12 + 8 \times 10 + 9 \times 5 + 10 \times 4}{2 + 7 + 12 + 10 + 5 + 4}$

$= \dfrac{301}{40} = 7.53$ (2 d.p.)

5 Mean $=$

$\dfrac{2 \times 3 + 6 \times 8 + 10 \times 19 + 14 \times 14 + 18 \times 6}{50}$

$= \dfrac{548}{50} = 10.96$

Median $= 8 + (25.5 - 11)\dfrac{4}{19} = 11.1$

6 (a) Put $y = x - 235$

Then we have

y	0–5	6–15	16–25	26–30
Freq	12	40	45	8

$\bar{y} = \dfrac{3 \times 12 + 11 \times 40 + 21 \times 45 + 28.5 \times 8}{12 + 40 + 45 + 8}$

$= \dfrac{1649}{105} = 15.7$

Now since $\bar{y} = \bar{x} - 235$, we have

$\bar{x} = \bar{y} + 235 = 250.7$

(b) Put $y = \dfrac{x - 35\,000}{10}$

Then y consists of

{ 735, 189, 2635, 1840, 1361 }

So $\bar{y} = \dfrac{6760}{5} = 1352$

Now since $\bar{y} = \dfrac{\bar{x} - 35\,000}{10}$, we have

$\bar{x} = 10\bar{y} + 35\,000$

$= 13\,520 + 35\,000 = 48\,520$

7 (a) $\bar{x} + a$

(b) $b\bar{x}$

8 ~390

9 ~3.2 kg

10 Mean *must* go down.

Median and mode might stay the same.

11 (a) • Length of internal > 200 000 unspecified.

• Awkward scale on x axis.

• difficult to accurately represent frequency densities on the y-axis.

(b) Median ~40 000.

Section 4

1 (a) (i) $\bar{x} = \dfrac{55}{5} = 11$

So the deviations are –9, –4, 1, 2, 10

$\Rightarrow s^2 = \dfrac{1}{n} \Sigma (x - \bar{x})^2$

$= \dfrac{1}{5}(81 + 16 + 1 + 4 + 100) = \dfrac{202}{5} = 40.4$

$\Rightarrow s = 6.36$ (2 d.p.)

(ii) $\Sigma x^2 = 4 + 49 + 144 + 169 + 441 = 807$

$\Rightarrow s^2 = \dfrac{807}{5} - 11^2 = 40.4$

$\Rightarrow s = 6.36$ (2 d.p.)

(b) We shall use a table which will supply the columns needed for both methods.

x	f	fx	$x - \bar{x}$	$(x-\bar{x})^2$	$f(x-\bar{x})^2$	fx^2 ($= xfx$)
4	3	12	–2	4	12	48
5	8	40	–1	1	8	200
6	12	72	0	0	0	432
7	10	70	1	1	10	490
8	2	16	2	4	8	128
	35	210			38	1298

(i) $\bar{x} = \dfrac{210}{35} = 6$

$s^2 = \dfrac{38}{35} = 1.09 \Rightarrow s = 1.04$ (2 d.p.)

(ii) $s^2 = \dfrac{1298}{35} - 6^2 = 1.09$

$\Rightarrow s = 1.04$ (2 d.p.)

2 $s = 6.58$ cm

3

x	f	mid-interval x	fx	$fx^2 \ (= xfx)$
5–	4	7.5	30	225
10–	9	12.5	112.5	1406.25
15–	15	17.5	262.5	4593.75
20–	13	22.5	292.5	6581.25
25–	7	27.5	192.5	5293.75
30–	3	32.5	97.5	3168.75
	51		987.5	21268.75

$$\bar{x} = \frac{987.5}{51} = 19.36 \ (2 \ d.p.)$$

$$s^2 = \frac{21268.75}{51} - 19.36^2 = 42.12$$

4 (a) $\bar{x}_A = \frac{550}{10} = 55 \ kg$

$$s^2_A = \frac{32\,500}{10} - 55^2 = 225 \ kg^2$$

$$\bar{x}_B = \frac{970}{15} = 64.67 kg$$

$$s^2_B = \frac{63\,500}{15} - 64.67^2 = 51.56 \ kg^2$$

(b) $\bar{x} = \frac{550 + 970}{10 + 15} = 60.8 \ kg$

$$s^2 = \frac{32\,500 + 63\,500}{10 + 15} - 60.8^2 = 143.36 \ kg^2$$

5 Advantage: easier to discern shape, centre and spread
Disadvantage: loss of accuracy and detail.

Median	27–28 cm
Upper quartile	31–32 cm
Lower quartile	23–24 cm
IQR	7 to 9 cm

Calculation gives:
Median = 27.6 cm
IQR = 8.8 cm
Median for spruce > Median for larch
IQR's are similar
Range is greater for larch than spruce

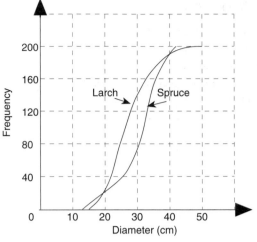

6 (a) $178 - 149 = 29$

(b) (i) lower quartile = 8.54 to 8.56

(ii) 81st percentile = 11.35 to 11.55

7 (a)

0	2 3 3 3 4 4 4 5 5 5 5 5 5 6 6 7 8 8 8 9
1	0 0 0 2 3 3 3 3 4 4 4 4 5 7 7 9
2	1 2 2 3 3 3 3 3 4 6
3	3 4 4 6
4	1 3

(b) Positively skewed

(c) $\bar{x} = \frac{738}{50} = 14.76$

$$s^2 = \frac{16\,526}{50} - 14.76^2 = 112.66$$

(d) Accuracy is reduced by grouping since detail is lost once all data in an interval is assumed to be concentrated at the centre of the interval.

8 (a) Skewness will be positive if the median is nearer to lower quartile than upper quartile and negative for vice versa.
Equidistant $\Rightarrow$ symmetrical distribution.

(b) $\sum f = 180$

Mass < 39 $\sum f = 37$
Mass < 44 $\sum f = 71$

Hence lower quartile in interval 40–44

Linear interpolation gives:

$$39.5 + \frac{8}{34} \times 5 \approx 40.7$$

Mass < 44 $\sum f = 71$
Mass < 49 $\sum f = 110$

$\Rightarrow$ Median in interval 45–49

Linear interpolation gives:

$$44.5 + \frac{19}{39} \times 5 = 46.9$$

Mass < 49 $\sum f = 110$
Mass < 59 $\sum f = 152$

$\Rightarrow$ upper quartile in interval 50–59

Linear interpolation gives:

$49.5 + \dfrac{25}{42} \times 10 = 55.5$

Upper quartile – median = 8.5

Median – lower quartile = 6.3

$\Rightarrow$ positively skewed.

(c)   The box and whisker plot looks like this:

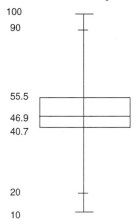

9   (a)

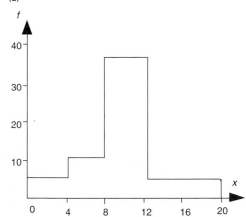

(Note that 4 is plotted for the range
   12–20.) The modal class is 8–12.

(b)

Ends	4	8	12	20
Cumulative frequencies	5	16	52	60

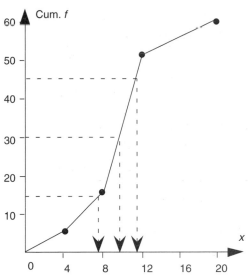

$\therefore$ Median = 9.6

Also $Q_1 = 7.6$ and $Q_3 = 11.2$

$\therefore$ Semi-interquartile range $= \dfrac{Q_3 - Q_1}{2}$

$= \dfrac{11.2 - 7.6}{2} = 1.8$

(c) $\bar{x} = 9.4$ and $s = 3.5646$

10

Ends	4.5	7.5	10.5	13.5
Cumulative frequencies	3	10	22	24

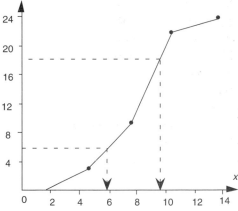

Interquartile range $= Q_3 - Q_1$

$= 9.5 - 5.8 = 3.7$

Section 5

1

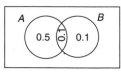

(a) $P(A' \cap B) = 0.1$

(b) P(A or B but not both) = 0.5 + 0.1 = 0.6

2 At least one Head $= \dfrac{7}{8}$

Score of more than $4 = \dfrac{2}{6}$

Independent $\Rightarrow \dfrac{7}{8} \times \dfrac{2}{6} = \dfrac{7}{24}$

3 3R, 4W, 5B

(a) possibilities are RRR or WWW or BBB

P(all same colour)

$= \left(\dfrac{3}{12}\right)^3 + \left(\dfrac{4}{12}\right)^3 + \left(\dfrac{5}{12}\right)^3$

$= 0.125 = \dfrac{1}{8}$

(b) possibilities are RWB or RBW or BRW or BWR or WBR or WRB.

P(all different) $= \dfrac{3}{12} \times \dfrac{4}{11} \times \dfrac{5}{10} + 5$ terms with the same numbers

$= 6 \times \dfrac{3}{12} \times \dfrac{4}{11} \times \dfrac{5}{10} = \dfrac{3}{11}$

4

$\left(\text{3R 4B}\right) \qquad \left(\text{5R 2B}\right)$

After transfer container 1 could be

2R 4B with probability $\dfrac{3}{7}$

or   3R 3B with probability $\dfrac{4}{7}$

P(red selected from 1 after transfer)

$= \dfrac{3}{7} \times \dfrac{2}{6} + \dfrac{4}{7} \times \dfrac{3}{6} = \dfrac{3}{7}$

5 Number of permutations of 3 out of 5

$= {}^5P_3 = \dfrac{5!}{2!} = 60$

Looking at the ratio of odds to evens in the list $\dfrac{3}{5} \times 60$

will be odd

$= 36$

6 4F 2M

Committee consists of 1 male, 2 females

1 male can be selected in 2 ways

2 females can be selected in ${}^4C_2 = 6$ ways

Hence 2 × 6 = 12 committees

7 3Y, 5G, 4R

(a) possibilities are GG, RR, GR, RG

P(no yellows in 1st two)

$= \dfrac{5}{12} \times \dfrac{4}{11} + \dfrac{4}{12} \times \dfrac{3}{11} + \dfrac{5}{12} \times \dfrac{4}{11} + \dfrac{4}{12} \times \dfrac{5}{11}$

$= \dfrac{72}{132} = \dfrac{6}{11}$

(b) Possibilities are YY, YR, RY, YG, GY

which is the complement of (a) hence $\dfrac{5}{11}$

(c) Let   A = 'fourth sweet is yellow'

B = 'first two sweets are red'

then $P(A \mid B) = \dfrac{P(A \cap B)}{P(B)}$

$A \cap B$ is RRYY or RRRY or RRGY

$P(A \cap B) = \left(\dfrac{4}{12} \times \dfrac{3}{11} \times \dfrac{3}{10} \times \dfrac{2}{9}\right)$

$+ \left(\dfrac{4}{12} \times \dfrac{3}{11} \times \dfrac{2}{10} \times \dfrac{3}{9}\right) + \left(\dfrac{4}{12} \times \dfrac{3}{11} \times \dfrac{5}{10} \times \dfrac{3}{9}\right)$

$= \dfrac{3}{110}$

B is RR

$P(B) = \dfrac{4}{12} \times \dfrac{3}{11} = \dfrac{1}{11}$

$P(A \mid B) = \dfrac{\frac{3}{110}}{\frac{1}{11}} = \dfrac{3}{10}$

8 $P(A \cup B) = P(A) + P(B) - P(A \cap B)$

$= P(A) + P(B) - P(A) \times P(B)$

since A, B are independent

$\Rightarrow \beta = \alpha + P(B) - (\alpha\, P(B))$

$\Rightarrow \beta - \alpha = P(B)(1 - \alpha)$

$\Rightarrow P(B) = \dfrac{\beta - \alpha}{1 - \alpha}$

9

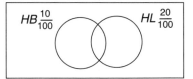

$P(HB \cup HL) = \dfrac{25}{100}$

$P(HB \cup HL) = P(HB) + P(HL) - P(HB \cap HL)$

$\Rightarrow \dfrac{25}{100} = \dfrac{10}{100} + \dfrac{20}{100} - P(HB \cap HL)$

$\Rightarrow P(HB \cap HL) = \dfrac{5}{100} = \dfrac{1}{20}$

$P(HL \mid HB) = \dfrac{P(HL \cap HB)}{P(HB)} = \dfrac{\frac{1}{20}}{\frac{10}{100}} = \dfrac{1}{2}$

10 (a) $P(A \mid B) = \dfrac{P(A \cap B)}{P(B)}$

$= \dfrac{P(A)\,P(B)}{P(B)} = P(A)$

since A, B are independent

$\Rightarrow P(A \mid B) = 0.2$

(b) $P(A \cap B) = P(A)\,P(B)$

$= 0.2 \times 0.15 = 0.03$

(c) $P(A \cup B) = P(A) + P(B) - P(A \cap B)$

$= 0.2 + 0.15 - 0.03 = 0.32$

11 $P(A \text{ wins}) = \dfrac{2}{3}$

A wins 3 games or 4 games out of 4 games

is $4 \left(\dfrac{2}{3}\right)^3 \left(\dfrac{1}{3}\right) + \left(\dfrac{2}{3}\right)^4 = \dfrac{16}{27}$

12 Let $\quad M = $ mice $\quad V = $ vole $\quad R = $ rest

$P(M) = \dfrac{1}{2} \qquad P(V) = \dfrac{1}{5}$

$P(R) = 1 - \dfrac{1}{2} - \dfrac{1}{5} = \dfrac{3}{10}$

$A = $ Albert, $B = $ Belinda, $K = $ Khalid,
$P = $ Poon

$P(A) = \dfrac{20}{100} \qquad P(B) = \dfrac{45}{100}$

$P(K) = \dfrac{10}{100} \qquad P(P) = \dfrac{25}{100}$

(a) (i) $P(A \cap M) = P(A) \cdot P(M)$
since independent

$= \dfrac{20}{100} \times \dfrac{1}{2} = \dfrac{1}{10}$

(ii) $P(A' \cap R) = \dfrac{80}{100} \times \dfrac{3}{10} = \dfrac{6}{25}$

(b) Using $P(M \mid B) - \dfrac{P(M \cap B)}{P(B)}$

we have $P(M \cap B) = P(B) \cdot P(M \mid B)$

$= \dfrac{45}{100} \times \dfrac{1}{3} = \dfrac{3}{20}$

(c) $P(M \mid K) = \dfrac{P(M \cap K)}{P(K)} = \dfrac{\frac{5}{100}}{\frac{10}{100}} = \dfrac{1}{2}$

(d) $P(M) = P(M \cap A) + P(M \cap B)$
$\qquad\qquad + P(M \cap K) + P(M \cap P)$

$= \dfrac{1}{10} + \dfrac{3}{20} + \dfrac{5}{100} + \dfrac{2}{10} = \dfrac{1}{2}$

(e) $P(B \mid M) = \dfrac{P(B \cap M)}{P(M)} = \dfrac{\frac{15}{100}}{\frac{1}{2}} = \dfrac{3}{10}$

Section 6

1 (a) See the graph below.

(b) $y = 9.789 + 0.924\,x$

(c) (i) $y = 15.98$ (ii) $y = 19.68$

The first is probably reliable, but not the second as it lies outside the interval for
x-values.

(d) Modify (c)(ii) to approx $y = 18.3$.

Appears to be levelling off to constant value.

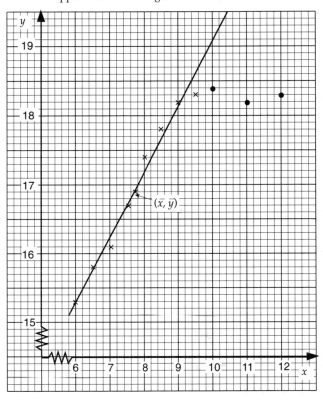

Section 6 solutions continued on p. 134

2 (a) See graph on the right.

(b) $y = -0.14 + 0.39x$

(c) $x = 60$, $y = 23.24$ Unsafe prediction since outside range of x-values

(d) (i) $y = -0.14 + 23.2t$

(ii) $z = -0.14 + 0.39x + 273$

(e) x is the independent or explanatory variable. Regression of x on y.

3 (a) (i) Yes

(ii) No (by substituting some values)

(b) Calculation of correlation coefficient, draw scatter diagram, x is the explanatory variable

$y = 107.91 - 1.49x$
$\therefore$ 70.7%

4 (a) $y = 3.67 + 0.038x$

(b) If no water applied, 3.67 will be the yield.

For every extra cm of water, an additional 0.038 tonnes will be gained.

(c) $x = 28$ gives $y = 4.72$, reliable since within the range of x-values given (30 to 120).

$x = 150$ gives $y = 9.33$, probably unreliable, since it is outside this interval.

5 (a) $y = 0.16 + 0.79x$

(b) A hen consumes 0.79 kg of food each week

(c) £11.43

6 (b) $y = 180.5 - 1.25x$

(c) b = yield per density, a = yield when density is zero. No sensible interpretation of a therefore.

(d) 144.32. The value of the density when yield becomes zero

(e) See graph on the right. Not adequate model, yx = constant is more plausible.

7 $r = 0.694$

One point (18, 32) is an outlier. Use all points except this one.

$b = 0.75$

$a = 5.31$

gives $y = 5.31 + 0.75x$

The interpretation of $b = 0.75$ is that every minute of advertising produced $0.75 \times 100 = 75$ sales.

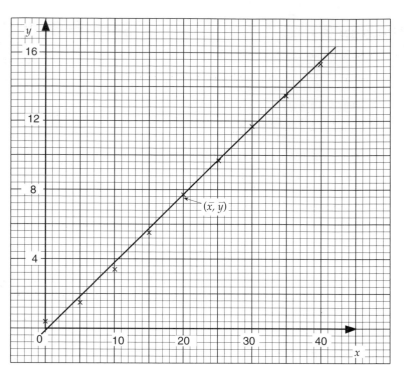

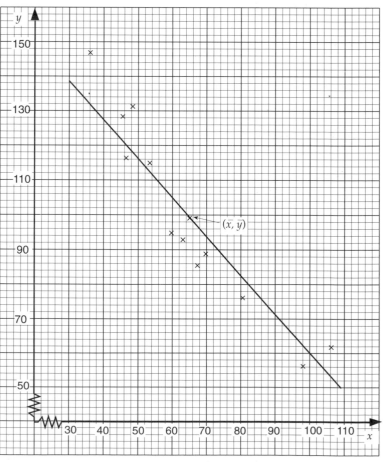

Section 7

1 Possible outcomes for X are 0, 1, 2, 3

$X = 0$ occurs in 1 way $\overline{R}\,\overline{R}\,\overline{R}$ with probability $\dfrac{5}{10} \times \dfrac{4}{9} \times \dfrac{3}{8}$

$X = 1$ occurs in 3 ways $\overline{R}\,\overline{R}R$ or $\overline{R}R\overline{R}$ or $R\overline{R}\,\overline{R}$ with probability $3 \times \dfrac{5}{10} \times \dfrac{5}{9} \times \dfrac{4}{8}$

$X = 2$ occurs in 3 ways $RR\overline{R}$ or $R\overline{R}R$ or $\overline{R}RR$ with probability $3 \times \dfrac{5}{10} \times \dfrac{4}{9} \times \dfrac{5}{8}$

$X = 3$ occurs in 1 way RRR with probability $\dfrac{5}{10} \times \dfrac{4}{9} \times \dfrac{3}{8}$

giving

x	0	1	2	3
$P(X = x)$	$\dfrac{60}{720}$	$\dfrac{300}{720}$	$\dfrac{300}{720}$	$\dfrac{60}{720}$

$P(X > 1) = \dfrac{300 + 60}{720} = \dfrac{1}{2}$

2 $P(X = 0) = P(TTT) = \dfrac{1}{2} \times \dfrac{1}{2} \times \dfrac{1}{2} = \dfrac{1}{8}$

$P(X = 1) = P(HTT \text{ or } THT \text{ or } TTH)$
$= 3 \times \dfrac{1}{2} \times \dfrac{1}{2} \times \dfrac{1}{2} = \dfrac{3}{8}$

$P(X = 2) = P(HHT \text{ or } HTH \text{ or } THH)$
$= 3 \times \dfrac{1}{2} \times \dfrac{1}{2} \times \dfrac{1}{2} = \dfrac{3}{8}$

$P(X = 3) = P(HHH) = \dfrac{1}{2} \times \dfrac{1}{2} \times \dfrac{1}{2} = \dfrac{1}{8}$

X can take the values 0, 1, 2, 3, so it is a discrete variable.

Also $\sum p_i = \dfrac{1}{8} + \dfrac{3}{8} + \dfrac{3}{8} + \dfrac{1}{8} = 1$,
so X satisfies the conditions of a discrete random variable.

3 We have $k + \dfrac{k}{2} + \dfrac{l}{3} + \dfrac{l}{4} = 1$

$\Rightarrow \dfrac{3k}{2} + \dfrac{7l}{12} = 1 \Rightarrow 18k + 7l = 12$ \qquad (1)

$P(X \le 2) = 2P(X > 2)$

$\Rightarrow \dfrac{3k}{2} = 2 \times \dfrac{7l}{12} \Rightarrow 9k = 7l$ \qquad (2)

Substituting gives
$18k + 9k = 12$
$\Rightarrow 27k = 12 \Rightarrow k = \dfrac{4}{9}$

$P(X = 2) = \dfrac{2}{9}$

4 $E(X) = 0 \times \dfrac{1}{8} + 1 \times \dfrac{3}{8} + 2 \times \dfrac{3}{8} + 3 \times \dfrac{1}{8} = \dfrac{12}{8} = 1.5$

$E(X^2) = 0^2 \times \dfrac{1}{8} + 1^2 \times \dfrac{3}{8} + 2^2 \times \dfrac{3}{8} + 3^2 \times \dfrac{1}{8} = \dfrac{24}{8} = 3$

$Var(X) = 3 - \left(\dfrac{3}{2}\right)^2 = \dfrac{3}{4}$

5 (a) $p(1) = k$
$p(2) = 4k$
$p(3) = 9k$

$9k + 4k + k = 1 \Rightarrow k = \dfrac{1}{14}$

(b) $E(X) = 1 \times \dfrac{1}{14} + 2 \times \dfrac{4}{14} + 3 \times \dfrac{9}{14} = \dfrac{18}{7}$

(c) $E(X^2) = 1^2 \times \dfrac{1}{14} + 2^2 \times \dfrac{4}{14} + 3^2 \times \dfrac{9}{14} = \dfrac{98}{14}$

$Var(X) = \dfrac{98}{14} - \left(\dfrac{18}{7}\right)^2 = 0.388$

6 18, 7, 23

7 20, 5, 20

8 $n = 27, \dfrac{7}{27}, 60\dfrac{2}{3}$

Section 8

1 (a) $z = \dfrac{87 - 80}{4} = 1.75$

Ans: 0.9599 (or 95.99%)

(b) $z = \dfrac{86 - 80}{4} = 1.5$

Ans: 0.0668 (or 6.68%)

(c) $z_1 = \dfrac{75 - 80}{4} = -1.25$
which gives 0.8944

$z_2 = \dfrac{78 - 80}{4} = -0.5$
which gives 0.6915

Subtract these areas to get
Ans: 0.2029 (or 20.29%)

(d) $z_1 = \dfrac{85.5 - 80}{4} = 1.375$
which gives 0.9154

$z_2 = \dfrac{77 - 80}{4} = -0.75$
which gives 0.7734

$\therefore$ required area $= 0.4154 + 0.2734$
$= 0.6888$ (or 68.88%)

(e) $z = \dfrac{82 - 80}{3} = 0.667$

Now 0.66 gives 0.7454
and 0.67 gives 0.7486
$\therefore$ 0.667 gives 0.7476
(We want $\dfrac{7}{10}$ of the way along between 54 and 86.
Now $86 - 54 = 32$ and
$\dfrac{7}{10} \times 32 = 22$ (approx)
$\therefore$ We want $54 + 22 = 76$.)
$\therefore$ Ans: 0.2476 (or 24.76%)

(f) $z = \dfrac{87 - 80}{6} = 1.167$

Now 1.16 gives 0.8770

and 1.17 gives 0.8790

$\therefore$ 1.167 gives 0.8784

$\therefore$ Ans: 0.1216 (or 12.16%)

2 (a) $\dfrac{x - 80}{4} = 1.175 \qquad \therefore x = 84.7$

(b) $\dfrac{x - 80}{4} = -0.44$ (note negative sign)

$$\therefore x = 78.24$$

(c) $\dfrac{x - 80}{4} = 0.841 \qquad \therefore x = 83.36$

(d) $\dfrac{x - 80}{4} = -1.037 \qquad \therefore x = 75.85$

3 (a) $\dfrac{32 - \mu}{5} = 0.26$

and solving gives $\mu = 30.7$

(b) $\dfrac{-3}{\sigma} = -0.53 \therefore \sigma = 5.66$

(c) We have $\dfrac{22 - \mu}{\sigma} = 0.64$

or $22 - \mu = 0.64\sigma$ [1]

and $\dfrac{17 - \mu}{\sigma} = -0.77$

$\Rightarrow 17 - \mu = -0.77\sigma$ [2]

[1] − [2] $\Rightarrow 5 = 1.41\sigma$

$\Rightarrow \sigma = 3.5\ 5$

$\Rightarrow \sigma^2 = 12.6$

Substituting into [1] gives $\mu = 22 - 2.27 \Rightarrow \mu = 19.7$

4 $\dfrac{40 - \mu}{\sigma} = -1.3$ and $\dfrac{48 - \mu}{\sigma} = 2.1$

$\therefore \mu = 43.06$ (2 d.p)

$\sigma = 2.35 \ldots$

$\therefore$ Variance = 5.54 (2 d.p)

5 $\dfrac{250 - \mu}{1.5} = -1.645$

$\therefore \mu = 252.5$ (1 d.p)

Appendix 1: Summary of basic set theory

For those students unfamiliar with the fundamental ideas and concepts of set theory, these are summarised below.

Elements of sets

A set is a collection of objects with some property in common and is a well-defined entity, in the sense that it is possible to say unambiguously whether something is in the set or not.

For example:

A = { the set of prime numbers less than 30 }

is a well-defined collection of positive integers, and, for example:

$7 \in A \qquad 39 \notin A \qquad 68.4 \notin A$

where the symbol $\in$ means 'is a member of' or 'is an element of' and $\notin$ means the negation of this, i.e. 'is not a member of'.

Sets are placed in curly brackets conventionally, as in the example above. They may be defined by a property, as in A above, or they may be defined by a list. For example:

B = { 1, 2, 3, 4, 5, 6 }

is the set of outcomes for an ordinary cubical die.

Sets may have an infinite number of members, e.g.

$\mathbb{Z}$, the **set of integers**

or $\mathbb{R}$, the **set of real numbers**.

All x { x : }

An alternative way in which sets may be defined by a rule is as in the example:

$C = \{ x : x \in \mathbb{Z}, x > \sqrt{3} \,)$

which should be read as:

'all x such that x is an integer ($\in \mathbb{Z}$) and x is bigger than $\sqrt{3}$.'

This could equivalently be defined by the list:

C = { 2, 3, 4, ... }

where the ... has an obvious meaning.

n(X)

The notation n(X) means the **number of elements** or objects in X and from the above examples:

n(A) = 10

since A = { 2, 3, 5, 7, 11, 13, 17, 19, 23, 29 }

n(B) = 6

n(C) is infinite

Subsets

An important relationship between sets is that of a **subset**:

$X \subset Y$ (read as 'X is a subset of Y')

if everything in X is also in Y. So, for example:

if A = { 1, 2, 3 }

and B = { 4, 1, 2, 3, }

then $A \subset B$

but $B \not\subset A$ i.e. B is not a subset of A.

For some sets that you may have met in pure maths, the following relationships hold:

$\mathbb{N} \subset \mathbb{Z} \subset \mathbb{R} \subset \mathbb{C}$

i.e. natural numbers are included in integers which are included in real numbers which are included in complex numbers.

Note that according to the definition,

$X \subset X$ always

since everything in X is automatically in X, and

$\varnothing \subset X$

since there are no things in $\varnothing$ which could not be in X.

The universal set $\mathcal{E}$

Any discussion or argument usually takes place within a context and in set theory, the context is called the universal set ($\mathcal{E}$). Once the universal set has been fixed, then all subsequent discussion takes place within that context. The universal set is analagous to the sample space.

An an example, suppose:

$\mathcal{E}$ = { positive integers }

and A = { prime numbers }

then we can define a new set, the **complement** of A written A' which is everything in $\mathcal{E}$ but not in A, which in this case will be non-prime (or composite) positive integers. For example:

$$7 \in A$$

and $\quad 18 \in A'$

Note, however, that $25.2 \notin A'$ as we had already confined the discussion to positive integers by defining $\mathcal{E}$.

Intersection of sets

The intersection of two sets A and B, written $A \cap B$ is defined formally by:

$$\{ x : x \in A \text{ and } x \in B \}$$

and so is the set of those things in common to both A and B.

Union of sets

The union of two sets A and B is written $A \cup B$ and is formally defined by:

$$\{ x : x \in A \text{ or } x \in B \}$$

and so is the set of those things which are either in A or B, or both.

Venn diagrams

The complement, union and intersection of sets can be neatly illustrated using **Venn Diagrams**. Conventionally the universal set is a rectangle and then sets within it are circles.

As an example consider:

$$\mathcal{E} = \{ 1, 2, 3, 4, 5, 6, 7, 8 \}$$
$$A = \{ 2, 4, 6, 8 \}$$
$$B = \{ 3, 4, 5, 6 \}$$

then the sets can be illustrated as follows:

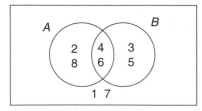

and it is clear from the diagram that for example:

$$A \cap B = \{ 4, 6 \}$$
$$A \cup B = \{ 2, 3, 4, 5, 6, 8 \}$$
$$A' = \{ 1, 7, 3, 5 \}$$
$$n(B') = 4$$
$$(A \cap B)' = \{ 1, 2, 3, 5, 7, 8 \}$$
$$(A \cap B) \subset A$$

Sets and set notation are widely used in probability theory and Venn Diagrams provide a useful problem-solving tool.

Appendix 2: The normal distribution function

The function tabulated below is $\Phi(z)$, defined as $\Phi(z) = \dfrac{1}{\sqrt{2\pi}} \displaystyle\int_{-\infty}^{z} e^{-\frac{1}{2}t^2}\, dt.$

z	$\Phi(z)$	z	$\Phi(z)$	z	$\Phi(z)$	z	$\Phi(z)$	z	$\Phi(z)$
0.00	0.5000	0.50	0.6915	1.00	0.8413	1.50	0.9332	2.00	0.9772
0.01	0.5040	0.51	0.6950	1.01	0.8438	1.51	0.9345	2.02	0.9783
0.02	0.5080	0.52	0.6985	1.02	0.8461	1.52	0.9357	2.04	0.9793
0.03	0.5120	0.53	0.7019	1.03	0.8485	1.53	0.9370	2.06	0.9803
0.04	0.5160	0.54	0.7054	1.04	0.8508	1.54	0.9382	2.08	0.9812
0.05	0.5199	0.55	0.7088	1.05	0.8531	1.55	0.9394	2.10	0.9821
0.06	0.5239	0.56	0.7123	1.06	0.8554	1.56	0.9406	2.12	0.9830
0.07	0.5279	0.57	0.7157	1.07	0.8577	1.57	0.9418	2.14	0.9838
0.08	0.5319	0.58	0.7190	1.08	0.8599	1.58	0.9429	2.16	0.9846
0.09	0.5359	0.59	0.7224	1.09	0.8621	1.59	0.9441	2.18	0.9854
0.10	0.5398	0.60	0.7257	1.10	0.8643	1.60	0.9452	2.20	0.9861
0.11	0.5438	0.61	0.7291	1.11	0.8665	1.61	0.9463	2.22	0.9868
0.12	0.5478	0.62	0.7324	1.12	0.8686	1.62	0.9474	2.24	0.9875
0.13	0.5517	0.63	0.7357	1.13	0.8708	1.63	0.9484	2.26	0.9881
0.14	0.5557	0.64	0.7389	1.14	0.8729	1.64	0.9495	2.28	0.9887
0.15	0.5596	0.65	0.7422	1.15	0.8749	1.65	0.9505	2.30	0.9893
0.16	0.5636	0.66	0.7454	1.16	0.8770	1.66	0.9515	2.32	0.9898
0.17	0.5675	0.67	0.7486	1.17	0.8790	1.67	0.9525	2.34	0.9904
0.18	0.5714	0.68	0.7517	1.18	0.8810	1.68	0.9535	2.36	0.9909
0.19	0.5753	0.69	0.7549	1.19	0.8830	1.69	0.9545	2.38	0.9913
0.20	0.5793	0.70	0.7580	1.20	0.8849	1.70	0.9554	2.40	0.9918
0.21	0.5832	0.71	0.7611	1.21	0.8869	1.71	0.9564	2.42	0.9922
0.22	0.5871	0.72	0.7642	1.22	0.8888	1.72	0.9573	2.44	0.9927
0.23	0.5910	0.73	0.7673	1.23	0.8907	1.73	0.9582	2.46	0.9931
0.24	0.5948	0.74	0.7704	1.24	0.8925	1.74	0.9591	2.48	0.9934
0.25	0.5987	0.75	0.7734	1.25	0.8944	1.75	0.9599	2.50	0.9938
0.26	0.6026	0.76	0.7764	1.26	0.8962	1.76	0.9608	2.55	0.9946
0.27	0.6064	0.77	0.7794	1.27	0.8980	1.77	0.9616	2.60	0.9953
0.28	0.6103	0.78	0.7823	1.28	0.8997	1.78	0.9625	2.65	0.9960
0.29	0.6141	0.79	0.7852	1.29	0.9015	1.79	0.9633	2.70	0.9965
0.30	0.6179	0.80	0.7881	1.30	0.9032	1.80	0.9641	2.75	0.9970
0.31	0.6217	0.81	0.7910	1.31	0.9049	1.81	0.9649	2.80	0.9974
0.32	0.6255	0.82	0.7939	1.32	0.9066	1.82	0.9656	2.85	0.9978
0.33	0.6293	0.83	0.7967	1.33	0.9082	1.83	0.9664	2.90	0.9981
0.34	0.6331	0.84	0.7995	1.34	0.9099	1.84	0.9671	2.95	0.9984
0.35	0.6368	0.85	0.8023	1.35	0.9115	1.85	0.9678	3.00	0.9987
0.36	0.6406	0.86	0.8051	1.36	0.9131	1.86	0.9686	3.05	0.9989
0.37	0.6443	0.87	0.8078	1.37	0.9147	1.87	0.9693	3.10	0.9990
0.38	0.6480	0.88	0.8106	1.38	0.9162	1.88	0.9699	3.15	0.9992
0.39	0.6517	0.89	0.8133	1.39	0.9177	1.89	0.9706	3.20	0.9993
0.40	0.6554	0.90	0.8159	1.40	0.9192	1.90	0.9713	3.25	0.9994
0.41	0.6591	0.91	0.8186	1.41	0.9207	1.91	0.9719	3.30	0.9995
0.42	0.6628	0.92	0.8212	1.42	0.9222	1.92	0.9726	3.35	0.9996
0.43	0.6664	0.93	0.8238	1.43	0.9236	1.93	0.9732	3.40	0.9997
0.44	0.6700	0.94	0.8264	1.44	0.9251	1.94	0.9738	3.50	0.9998
0.45	0.6736	0.95	0.8289	1.45	0.9265	1.95	0.9744	3.60	0.9998
0.46	0.6772	0.96	0.8315	1.46	0.9279	1.96	0.9750	3.70	0.9999
0.47	0.6808	0.97	0.8340	1.47	0.9292	1.97	0.9756	3.80	0.9999
0.48	0.6844	0.98	0.8365	1.48	0.9306	1.98	0.9761	3.90	1.0000
0.49	0.6879	0.99	0.8389	1.49	0.9319	1.99	0.9767	4.00	1.0000
0.50	0.6915	1.00	0.8413	1.50	0.9332	2.00	0.9772		

Appendix 3: Key Skills

You work on this book will provide opportunities for gathering evidence towards Key Skills, especially in Communication and Application of Number.

These opportunities are indicated by the 'key' icon, e.g. 🔑 **C** 3.2. This means that the exercise contains the type of task that is relevant to Communication Level 3 and may help you gather evidence specifically for C3.2.

The places where Key Skills references are given are listed below, together with some other ideas about possible opportunities for gathering evidence.

Communication

C3.1a   Contribute to a group discussion

You may have opportunities for group discussion, e.g. about how you collect and interpret data.

C3.2   Read and synthesise information from two extended documents that deal with a complex subject. One of these documents should include at least one image.

See pages 8, 21, 25, 38, 39, 53, 54, 83, 96, 97 and 99 of this book.

Your work on S1 provides lots of opportunities for reading and synthesising information, especially if you carry out any project that involves analysing tables or statistical diagrams.

Application of number

N3.1   Plan interpret information from two different types of sources, including a large data set.

N3.2   Carry out multi-stage calculations.

N3.3   Interpret results of your calculations, present your findings and justify your methods. You must include at least one graph, one chart and one diagram.

See pages 5 and 12 of this book.

Your work on S1 will provide a wide range of opportunities for covering the Application of Number Key Skills, especially if you carry out a project or extended assignment that involves collecting and interpreting data, and presenting your findings.